A Publication of the
National Research Council of Canada
Monograph Publishing Program

Philip J. Currie, Wann Langston, Jr., & Darren H. Tanke

A New Horned Dinosaur From an Upper Cretaceous Bone Bed in Alberta

with contributions from Patricia E. Ralrick, Ryan C. Ridgely, and Lawrence M. Witmer

NRC·CNRC
NRC Research Press

ISBN 978-0-660-19819-4
NRC No. 49729

Library and Archives Canada Cataloguing in Publication

Currie, Philip J., 1949-
A new horned dinosaur from an Upper Cretaceous bone bed in Alberta / Philip John Currie, Wann Langston, Jr., Darren H. Tanke.

Issued by: National Research Council Canada.
Includes bibliographical references.
ISBN 978-0-660-19819-4

1. Pachyrhinosaurus. 2. Paleontology—Alberta.
3. Dinosaurs—Alberta.
I. Langston, Wann, 1921- II. Tanke, Darren H III. National Research Council Canada IV. Title.

QE862.O65C87 2008 567.915 C2008-980235-7

The Publisher wishes to thank the County of Grande Prairie, AB with a special thanks to Walter Paszkowski for their financial contribution to this book.

Inquiries: Monograph Publishing Program, NRC Research Press, National Research Council of Canada, Ottawa, Ontario K1A 0R6, Canada.
Web site: http://pubs.nrc-cnrc.gc.ca

Correct citation for this publication: Currie, P.J., Langston, W., Jr., and Tanke, D.H. 2008. A New Horned Dinosaur from an Upper Cretaceous Bone Bed in Alberta. NRC Research Press, Ottawa, Ontario, Canada. 144p.

CONTENTS

Contributors

Philip J. Currie, University of Alberta, Department of Biological Sciences, CW405 Biological Sciences Building, Edmonton, AB T6G 2E9, Canada (e-mail: philip.currie@ualberta.ca).

Wann Langston, Jr., University of Texas at Austin, J.J. Pickle Research Campus, Vertebrate Paleontology Laboratory #6, 10100 Burnet Road, Austin, TX 78758, USA (e-mail: wannl@mail.utexas.edu).

Patricia E. Ralrick, Interdisciplinary Graduate Program, University of Calgary, 2500 University Drive NW, Calgary, AB T2N 1N4, Canada (e-mail: pattyralrick@hotmail.com).

Ryan C. Ridgely, Department of Biomedical Sciences, Ohio University College of Osteopathic Medicine, Athens, OH 45701, USA (e-mail: ridgely@ohio.edu).

Darren H. Tanke, Royal Tyrrell Museum of Palaeontology, Box 7500, Drumheller, AB T0J 0Y0, Canada (e-mail: darren.tanke@gov.ab.ca).

Lawrence M. Witmer, Department of Biomedical Sciences, Ohio University College of Osteopathic Medicine, Athens, OH 45701, USA (email: Witmer@oucom.ohiou.edu).

Foreword

I am unlikely to forget my first encounters with *Pachyrhinosaurus lakustai*. During the late 1980s I was a young graduate student at the University of Toronto with a keen interest in horned dinosaurs. Each summer, I traveled to Alberta to work in the fossil wonderland known as Dinosaur Provincial Park. I also spent time studying fossils at the Tyrrell Museum of Palaeontology (not yet "Royal", at least not formally). Every year, Darren Tanke, whose passion for horned dinosaurs seemed boundless, would take me into the collections and pull out drawer after drawer of bones belonging to this strange new beast from Pipestone Creek. I was amazed at the sheer volume of the bone bed assemblage. Most dinosaur species are known from one or two fragmentary specimens, yet here was a previously unknown animal represented by numerous individuals, with virtually every bone in the skeleton represented. A broad suite of ages was present too, from sheep-sized juveniles to massive adults larger than rhinos. Even then, I knew that I was looking at one of the best sampled dinosaurs on the planet.

Most remarkable of all was (and still is) the skull. Granted, ceratopsid dinosaurs have some of the most spectacular skulls known. The largest species, like *Triceratops*, achieved skull lengths of more than 3 meters, exceeding that of all other land-dwelling vertebrates past or present. Whereas a *Tyrannosaurus* skull looks in many respects like that of an oversized iguana, those of *Triceratops* and its horned kin are highly specialized, both unique and bizarre. The jaws include a toothless, parrot-like beak up front and hundreds of tightly appressed teeth behind, a formidable tear-and-slice combination apparently suited for consuming a high-fiber, low quality diet of plants. The nose—in particular the outermost chamber—is ridiculously expanded, dominating the front of the head. Directly above the brain is a sizeable bony cavity of uncertain function. Most impressive of all is the pair of skull roof bones (parietal and squamosal) that stretches rearward more than a metre to form an impressive shield-like frill. Add to this framework a trio of horns, one over the nose and one each over the eyes, and the result is a prehistoric marvel.

Pachyrhinosaurus lakustai, however, ups the bizarre quotient still further, adorning the skull with a bewildering array of hooks, horns spikes, bosses, bumps, and other unnamed excrescences of bone. From stem to stern, these osseous accoutrements include: knob-like growths, some pointy, on the tip of the snout; a broad and elongate platform, or boss, of bone above the enlarged nasal region; another pair of thick, roughened bosses above the orbits; a hornlet (epijugal) on the tip of each cheek bone (jugal); wavy undulations along the sides of the frill that were topped with accessory ossifications (epoccipitals); a pair of curved, laterally projecting spikes on the back of the frill; another, shorter pair of spikes (horns?) directed inward on the rear parietal margin; and variably developed spikes, horns, and bumps (I honestly don't know what to call them) pointing upward, unicorn-like, from the midline of the parietal frill. As if all of these bells and whistles weren't enough, the exceptional sample of *P. lakustai* from Pipestone Creek demonstrates that juveniles possessed small horns above the nose and eyes, as in most other ceratopsids, and that these protuberances were later transformed into robust, flattened bosses as the animals approached adult size. Then, apparently after reaching full maturity, adults frequently resorbed the central portions of these nasal and supraorbital bosses, leaving behind irregular bony craters that give the appearance of having been blowtorched. Exactly how these unusual structures were modified, let alone why, remains a mystery.

In a manner befitting such a magnificent beast, this volume is a landmark scientific contribution. Admittedly, it has been a long time in the making. One of the authors (Tanke) likes to note that he did not have any children when he began working on *P. lakustai*, whereas today, as this volume is published, his children are in college! To be fair, however, this project was a daunting task, made challenging by the embarrassment of fossil riches from the Pipestone Creek bone bed. Philip Currie, recognizing the exceptional potential offered by such a large fossil sample, decided early on that the paper should include not only element-by-element descriptions but also a discussion of growth-based changes. To complicate matters, new specimens came to light as the manuscript developed, resulting in layers of revisions along with additional figures.

Nevertheless, as someone who has been anticipating publication of this study for over two decades, the final result is certainly well worth the wait. The osteological descriptions herein comprise the most detailed treatment for any ceratopsid dinosaur, and one of the most comprehensive descriptive works to date for a dinosaur species. The lengthy and rigorous text presented in the volume's core contribution by Currie, Langston, and Tanke is complemented by nearly 50 exquisite illustrations by Rod Morgan and others. This paper will undoubtedly serve as a descriptive standard within dinosaur paleontology for years to come. The treatment of the brain cavity and inner ear by Witmer and Ridgley, richly illustrated on the basis of three-dimensional reconstructions derived from CT scanning, is an unexpected surprise. It is also a first for ceratopsids, and one of the few digital imaging studies of a dinosaur brain endocast. Finally, the brief taphonomic investigation by Ralrick and Tanke provides important evidence as to the genesis of the Pipestone Creek bone bed.

The broader significance of this volume is minimally two tiered. *Pachyrhinosaurus lakustai* joins a rapidly growing parade of ceratopsid dinosaurs. A 2004 review of the group by Peter Dodson, Cathy Forster, and I recognized 16 valid species named over a period of about one century. Remarkably, that number has doubled in less than five years, with several recently published taxa added to the mix and other descriptions nearing completion. Until recently, *Pachyrhinosaurus canadensis* was regarded as a thick-headed, hornless oddity within Ceratopsidae. Today the known diversity of pachyrhinosaur-like forms (i.e., with nasal and supraorbital bosses) has ballooned to at least five taxa—with a possible sixth taxon from the Dinosaur Park Formation—comprising a substantial portion of the total diversity of Centrosaurinae. At the lower, alpha taxonomic level, then, the osteological descriptions in these pages will immediately become the primary resource for those seeking to make comparisons with other short-frilled (centrosaurine) ceratopsids. This study will also be profitably mined for phylogenetic characters. Ultimately, it is my hope that the data presented in this volume, when incorporated into larger datasets, will add some phylogenetic resolution to this (still foggy) corner of Dinosauria.

The higher level significance of this book relates to the fact that ceratopsids provide arguably the best opportunity among dinosaurs to investigate the tempo and mode of evolution. The horned dinosaur radiation is particular amenable to study for several reasons, including an abundant fossil record, preservation of species-specific signaling structures (horns and frills), and restricted temporal and geographic distributions. Ceratopsids flourished for all but the very last of their tenure along the eastern coastline of a peninsular landmass (Laramidia) that resulted from incursion of the Cretaceous Western Interior Seaway. Despite the fact that this landmass was less than one-fifth the size of present day North America, growing evidence suggests that distinct species, and perhaps independent radiations, of these horned giants co-occurred in the northern and southern regions, respectively, of the WIB. For example, all pachyrhinosaur-like centrosaurines are currently restricted to the northern portion of this landmass (Alberta, Montana, and Alaska), whereas the few examples known from the south (mostly Utah) appear to share closer affinity with the basal, long-horned *Albertaceratops*. However, before a comprehensive analysis of this radiation can be undertaken, more baseline data are needed on specific taxa. Thanks to this contribution, *P. lakustai* is now the single largest datapoint within Ceratopsidae, providing a critical piece of the puzzle. Like any good science, this volume is not an end, but a beginning, enabling other kinds of studies to be done with greater rigor.

And finally, aside from its diverse contributions to science, *P. lakustai* without doubt ranks as one of the most bizarre dinosaurs known, and is purely and simply one of the most marvelous animals generated in almost 4 billion years of evolutionary history on Earth.

Scott D. Sampson, Ph.D.
Research Curator, Utah Museum of Natural History
Adjunct Research Associate Professor,
Department of Geology and Geophysics
University of Utah

1. A new species of *Pachyrhinosaurus* (Dinosauria, Ceratopsidae) from the Upper Cretaceous of Alberta, Canada

PHILIP J. CURRIE, WANN LANGSTON, JR., AND DARREN H. TANKE

Abstract

A densely packed bone bed near Grande Prairie, Alberta, has produced abundant remains of a new species of ceratopsid. A minimum number of 27 individuals is represented in the part of the bone bed that has been excavated so far. The new animal—*Pachyrhinosaurus lakustai* sp. nov.— is closely related to the centrosaurine *Pachyrhinosaurus canadensis*, which has been recovered from younger beds in southern Alberta. It differs from the geologically younger species in having a relatively shorter nasal boss that is well separated from the supraorbital bosses. Juveniles of the new taxon resemble juveniles of *Centrosaurus* and other centrosaurines. However, the cranial morphology underwent a remarkable ontogenetic change, in which the nasal and supraorbital horns of the juveniles transformed into a huge nasal boss and smaller supraorbital bosses, and the frill became adorned with spikes and horns on top of and at the back of the parietal. Although there is some indication that the species may have been sexually dimorphic at maturity, it is not possible to separate sexual variation from individual and ontogenetic variation without a much larger sample. It is quite clear that the nasal boss supported some sort of keratinous structure, although it is not possible to determine its shape and function. No cause has been determined for the apparent catastrophic demise of a herd of *P. lakustai.*

Key words: Dinosauria, Ceratopsia, *Pachyrhinosaurus lakustai*, Late Cretaceous, Alberta, Canada.

Résumé

Un riche dépôt ossifère près de Grande Prairie (Alberta) a mené à la découverte de nombreux restes d'une nouvelle espèce de cératopsidés. Au moins 27 spécimens se trouvent dans les ossements dégagés jusqu'à présent. Le nouvel animal — ***Pachyrhinosaurus lakustai*** sp. nov. — est étroitement lié à *Pachyrhinosaurus canadensis*, centrosauriné des dépôts plus jeunes du sud de l'Alberta. Il se distingue des espèces qui sont géologiquement plus jeunes par la présence d'un renflement (bourrelet) nasal plus court et bien démarqué des renflements supraorbitaux. Les jeunes de ce nouveau taxon ressemblent aux jeunes de *Centrosaurus* et des autres centrosaurinés. Toutefois, la morphologie crânienne a subi un important changement ontogénique grâce auquel les cornes nasale et supraorbitales des juvéniles se sont transformées en un énorme renflement nasal et en petits bourrelets supraorbitaux, et la collerette est maintenant recouverte d'épines et de cornes au-dessus et à l'arrière du pariétal. Bien que certains faits laissent croire que l'espèce est peut-être sexuellement dimorphe à l'âge adulte, il n'est pas possible de faire la distinction entre la variation sexuelle, individuelle et ontogénique sans augmenter la taille de l'échantillon. Il est très clair que le renflement nasal a servi d'appui à une structure cornée, même si sa forme et son rôle demeurent encore un mystère. Aucune raison apparente n'explique la fin catastrophique du troupeau de *P. lakustai.*

Mots-clés : Dinosauria, Ceratopsia, *Pachyrhinosaurus lakustai*, Crétacé tardif, Alberta, Canada.

Introduction

Brief history of pachyrhinosaur discoveries

More than 50 years ago, some unusual ceratopsian specimens were recovered from the St. Mary River Formation on the north side of the Little Bow River, Alberta, and from exposures 25 km away that are known as Scabby Butte (Fig. 1). Sternberg (1950) described this bizarre ceratopsian dinosaur as *Pachyrhinosaurus canadensis*. He placed his animal in a new family, the Pachyrhinosauridae. Unlike other ceratopsians with nasal horn cores, *P. canadensis* had an enormous, broad boss of bone above and behind the nares. Sternberg also noted that many of the bones (maxilla, jugal, quadrate, dentary, coronoid process, angular, surangular, and articular) were relatively shorter, deeper, and more massive, suggesting great muscular power. Staff of the National Museum of Canada (now the Canadian Museum of Nature) reopened the bone bed at Scabby Butte in 1957 (Langston 1975). The Little Bow River site and the bone bed at Scabby Butte have

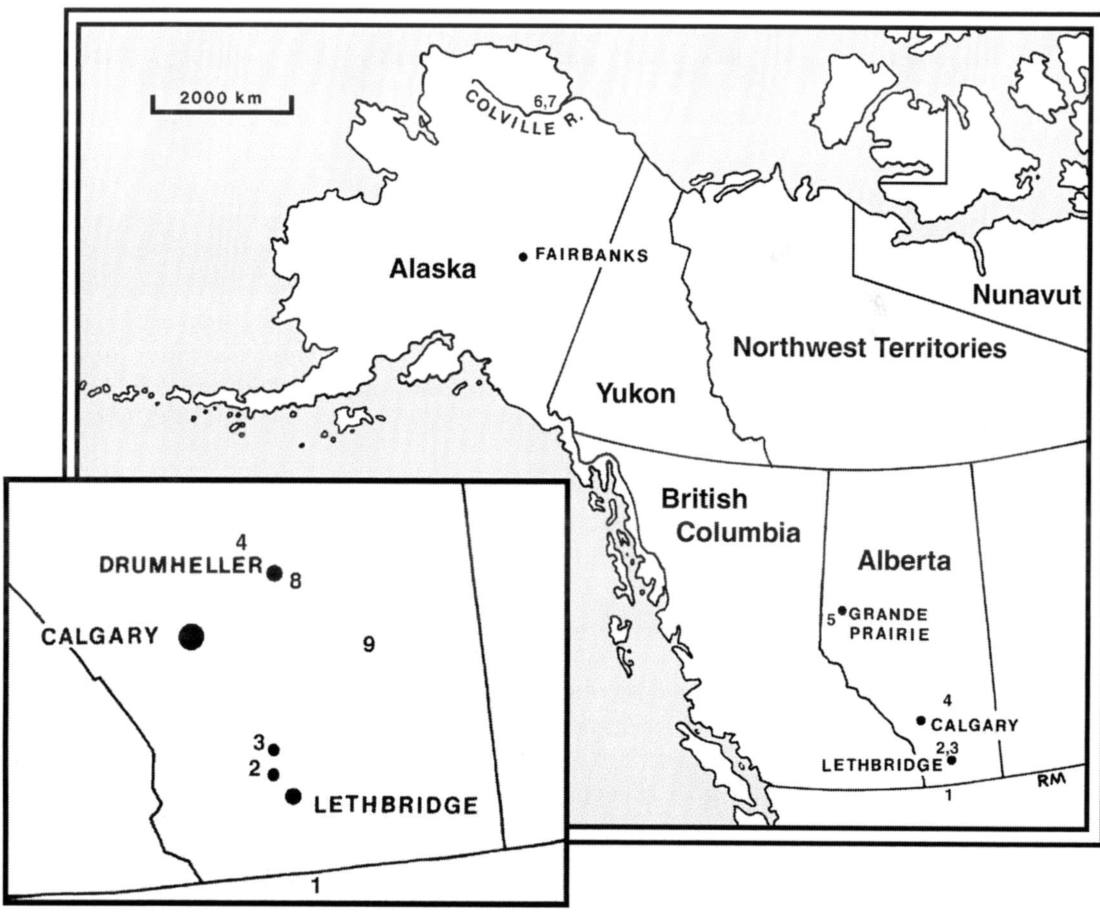

Fig. 1. Pachyrhinosaur localities. 1, Achelousaurus horneri *locality at Landslide Butte, Montana (Sampson 1995); 2,* Pachyrhinosaurus canadensis *referred specimens, St. Mary River Formation, Scabby Butte bone bed (Langston 1975); 3,* Pachyrhinosaurus canadensis *holotype and paratype, St. Mary River Formation, Little Bow River (Sternberg 1950); 4,* Pachyrhinosaurus canadensis *("The Drumheller specimen" of this paper), Horseshoe Canyon Formation, northwest of Drumheller (Langston 1967); 5,* Pachyrhinosaurus lakustai, *Wapiti Formation, Pipestone Creek and Wapiti River bone beds; 6,* Pachyrhinosaurus sp., *Prince Creek Formation, Colville River (Clemens and Nelms 1993); 7, Tegoseak bone bed, Colville River (May and Gangloff 1999); 8, Rosedale and Cambria (TMP 1999.63.12) sites southeast of Drumheller; 9, Dinosaur Provincial Park.*

produced nine partial skulls of *P. canadensis* (CMN 8860, CMN 8866, CMN 8867, CMN 9485, CMN 10645, CMN 10663, CMN 21863, CMN 21864, TMP 1982.52.1) and numerous postcranial remains (Sternberg 1950; Langston 1967, 1968, 1975).

Another *P. canadensis* skull was found near Drumheller (Langston 1967, 1968) in a Horseshoe Canyon Formation bone bed (Tyrrell Locality L1504) dominated by the hadrosaur *Edmontosaurus regalis*. The specimen has been on display in the Drumheller Dinosaur and Fossil Museum (now known as the Drumheller Valley Interpretive Centre) since 1967. Because the specimen does not have a catalogue number, it will be referred to as the Drumheller skull in this paper. The quarry that produced the Drumheller skull has been reopened several times by the Tyrrell Museum, and additional possible pachyrhinosaur bones (TMP 1993.29.4, TMP 1993.29.5, TMP 1995.91.1) have been recovered. The Horseshoe Canyon Formation in the Drumheller region has also produced an uncollected skull near Rosedale and a partial skull (nasal boss is catalogued as TMP 1999.63.12) near Cambria (Tanke 2006).

Mr. Al Lakusta, a science teacher from Grande Prairie, Alberta, discovered the Pipestone Creek bone bed in the fall of 1972 (Tanke 2006). He recovered a number of dinosaur bones of unknown affinity from the east bank of Pipestone Creek over the next 3 years. Prior to this discovery, only a waterworn hadrosaur pedal phalanx and fragmentary bones had been reported from the Campanian–Maastrichtian Wapiti Formation, so the discovery of a rich monodominant bone bed was surprising. P.J. Currie examined some of the Lakusta collection in 1979 and identified the material as ceratopsian. Currie and D.H. Tanke visited the monodominant bone bed along Pipestone Creek in June 1983 but were unable to identify what kind of centrosaurine was in the bone bed. In the summer of 1985, Tanke identified in the Grande Prairie Pioneer Museum a roughly prepared pachyrhinosaur cranial roof (TMP 1985.112.1) from the bone bed. The Lakusta collection was donated to the Royal Tyrrell Museum of Palaeontology in November 1985.

The Tyrrell Museum commenced a preliminary excavation at the locality in the summer of 1986 and used heavy equipment late that autumn to uncover a larger area (240 m^2). Subsequent work by the Museum was carried out in the summers of 1987, 1988, and 1989 by field crews consisting largely of volunteers and Grande Prairie Regional College staff. Preliminary stratigraphic interpretations were made by Gopol Dongal, a volunteer worker from Nepal, who participated in the project in the 1988 field season. The area of the bone bed excavated so far is approximately 52 m^2. The deposit has also been identified 120 m downstream on the west bank of Pipestone Creek (Fig. 1) where the snout region of an adult pachyrhinosaur skull (TMP 1987.55.285) and a few other bones were collected. Almost 2500 ceratopsian bones, including 15 skulls of varying completeness, have been recovered. A second pachyrhinosaur bone bed was subsequently found 32 km to the west on the north side of the Wapiti River (Fanti and Currie 2007). This site is stratigraphically higher, and the supraorbital bosses are smoother than those found at Pipestone Creek. It is conceivable that the Wapiti River specimens represent a different species, but further excavation and study is necessary. A partial pachyrhinosaur skull (TMP 2001.11.1) was recovered from this bone bed in 2001 but has not been prepared. At least seven more partial skulls, with nasal bosses, were counted at the site when a University of Alberta crew did a trial excavation in 2007.

Additional specimens were reported from west-central Alberta (Tanke 1988, 2006) and from northern Alaska (Nelms and Clemens 1989; Clemens and Nelms 1993; May and Gangloff 1999; Fiorillo 2004). An isolated ceratopsian quadrate from the Northwest Territories may also be from a pachyrhinosaur (Russell 1984). Another *Pachyrhinosaurus*-like animal—*Achelousaurus horneri* Sampson 1995—has been recovered from northern Montana. The skull and skeleton of a pachyrhinosaur (TMP 2002.76.1) were collected from Dinosaur Provincial Park. This animal appears to represent another new species (Ryan et al. 2006). All specimens attributed to pachyrhinosaurs are from Upper Cretaceous (Campanian–Maastrichtian) sediments.

The *P. canadensis* specimens recovered from the St. Mary River Formation at Scabby Butte were found in a bone bed dominated by this species and by hadrosaurs (Langston 1976). Although the nasal bosses show some variability that, in part, can be attributed to postmortem distortion, they are morphologically similar to each other and to the specimens collected along the Little Bow River. The Drumheller skull from the Horseshoe Canyon Formation (Langston 1967) has a deeper nasal boss, the lateral surface of which is deeply furrowed. Although it is from an equivalent time period, the nasal boss morphology of the Drumheller specimen is somewhat different than that of the topotype material of *P. canadensis*. Pending further analysis, the Drumheller skull is treated separately in the description and phylogenetic analysis. The Alaska skull falls even further outside the range of variation of the *P. canadensis* topotype series and may well represent a different genus and (or) species.

The Pipestone Creek bone bed

The Pipestone Creek bone bed is located about 22 km southwest of the city of Grande Prairie in west-central Alberta. The bone bed occurs in Upper Cretaceous rocks of the Wapiti Formation. The age of the Wapiti Formation sedimentary package is not known precisely. However, the unit is believed to be laterally equivalent to part or all of the upper Campanian (Judithian) to Maastrichtian (Lancian) formations exposed along the Red Deer River 580–700 km to the southeast. The stratigraphy, depositional environments and coal deposits of the Wapiti beds were investigated in the late 1980s and early 1990s by the Geological Survey of Canada and the Alberta Research Council. A survey map of stratigraphic sections studied by Dawson et al. (1989) shows the Pipestone quarry is positioned geographically between and stratigraphically close to their locality numbers 18 and 19. As the strata in this region are relatively flat lying, the Pipestone Creek site falls in the Red Willow Coal Measure section, which is dated palynologically as latest Campanian.

The monodominant bone bed (Eberth and Getty 2005) of Pipestone Creek (Fig. 1) is composed mostly of the remains of an animal similar to *P. canadensis*. The specimens shed additional light on the cranial anatomy of pachyrhinosaurs. An intriguing aspect of the collection is the presence of large numbers of bones from immature individuals (Tanke 1988; Sampson et al. 1997). A considerable range in individual size exists and reveals a striking series of ontogenetic changes unparalleled in other neoceratopsian assemblages. Moreover, the high variability in development of cranial ornamentation is greater than in any other ceratopsian presently known. These animals are distinct enough from *P. canadensis* and the pachyrhinosaur from Alaska to suggest that they represent a new species. There appear to be a minimum of three closely related centrosaurine species with nasal bosses that form a clade referred to in this paper as pachyrhinosaurs.

Based on the number of predentaries collected, the Pipestone Creek bone bed has yielded the remains of more than 27 individuals of the new species (Table 1). Large numbers of postcranial bones confirm the large number of individuals in the bone bed. There is a considerable range in size and ontogenetic development. The third author prepared much of the collection, and assembled two composite skeletons and four skulls (three adult and one juvenile) for display in the Royal Tyrrell Museum of Palaeontology, the Institute of Vertebrate Paleontology and Paleoanthropology (Beijing), and various traveling exhibits (Acorn 1993; Hamada 2001). A cast of a complete skeleton was also put on display at the Grande Prairie Regional College in 2002. This paper is limited mainly to a description of the skull and mandible of the new taxon and augments previous suggestions by Tanke (1988), Sampson and Tanke (1990), and Sampson et al. (1997) concerning secondary sexual characters and the remarkable ontogenetic changes in centrosaurines.

Table 1. Minimum number of individuals of *Pachyrhinosaurus lakustai* as determined by predentary bones.

Specimen No.	Ventral length (mm)	Age of animal
TMP 1983.55.72	Fragmentary	Adult
TMP 1985.112.103	Fragmentary	Adult
TMP 1985.112.114	Fragmentary	Subadult B
TMP 1986.55.30	195	Adult
TMP 1986.55.70	Fragmentary	Adult
TMP 1987.55.116	195e	Adult
TMP 1987.55.118	60e	Juvenile
TMP 1987.55.123	Fragmentary	Subadult A
TMP 1987.55.174	206	Adult
TMP 1987.55.219	210	Adult
TMP 1988.55.151	Fragmentary	Subadult B
TMP 1988.55.161	210	Adult
TMP 1988.55.182	150e	Subadult B
TMP 1988.55.191	Broken	Subadult B
TMP 1988.55.227	215e	Adult
TMP 1989.55.30	175	Subadult B
TMP 1989.55.107	160	Subadult B
TMP 1989.55.148	220	Adult
TMP 1989.55.185	175e	Adult
TMP 1989.55.304	77	Juvenile
TMP 1989.55.339	93e	Subadult A
TMP 1989.55.532	110	Subadult A
TMP 1989.55.581	101	Subadult B
TMP 1989.55.1223	221	Adult
TMP 1989.55.1290	202	Adult
TMP 1989.55.1398	200	Adult
TMP 1989.55.1500	130	Subadult A
TMP 2002.76.1	240	Adult

Note: Length measurement is from the anterior tip of the predentary to the anterior end of the midline concavity for contact with the dentary. Estimated lengths were only done for those predentaries that were complete enough to restore the anterior tip with confidence. "Subadult A" is an arbitrary designation for animals that are smaller and younger than "subadult B." e, estimate.

A volcanic ash sample from the vicinity of the bone bed yielded a radiometric date of 73.27 million years ago ± 250 000 years (D. Eberth, personal communication). This makes the sediments younger than any of the Campanian sediments exposed in Dinosaur Provincial Park (Eberth 2005), where another pachyrhinosaur skeleton

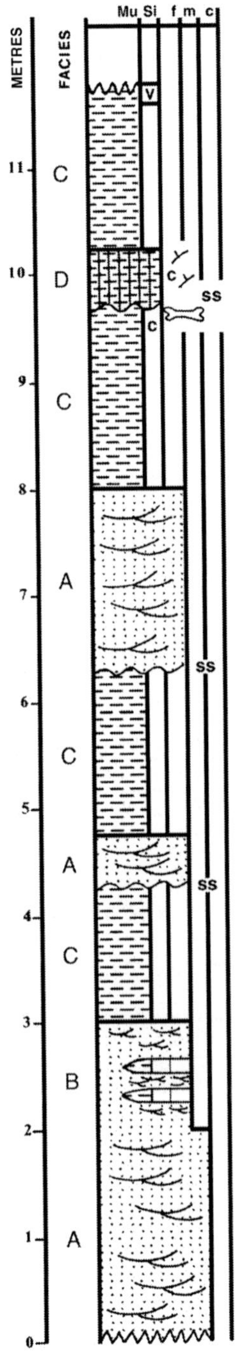

Fig. 2. Columnar section at the Pipestone Creek bone bed site. Column headings are as follows: Mu, mudstone; Si, siltstone; f, m, c, fine-, medium-, and coarse-grained sandstones. The symbolic bone represents the bone bed layer. c, small coal lenses; v, volcanic ashes, sandstone.

has been recovered recently (Ryan et al. 2006). Sediment samples taken from the bone bed, including one removed from a braincase (TMP 1989.55.925), showed relatively poor recovery of palynomorphs. Of the 11 species identified, most have relatively long age ranges (D.R. Braman, Royal Tyrrell Museum of Palaeontology, Drumheller, Alta., personal communication). One exception is *Wodehouseia edmontonicola*, which is restricted to the interval between the upper Bearpaw Formation and Coal Seam 9 of the Horseshoe Canyon Formation along the Red Deer River of southern Alberta. This agrees with the radiometric date suggesting that the bone bed is late Campanian in age. Additional palynological samples from various levels on the Wapiti River upstream and downstream from the mouth of Pipestone Creek are consistent with a Late Campanian age for this section of the Wapiti Formation (Dennis Braman, unpublished reports on file at the Tyrrell Museum).

The exposed section containing the bone bed consists of some 12 m of alternating soft shale and hard, fine-grained white sandstone and siltstone arranged in fining-upward couplets that are suggestive of a meandering river system (Fig. 2). The bones occur within a relatively soft dark carbonaceous siltstone that is at least 4 m thick. Typical of mudstone deposits, there is extreme plastic distortion of many of the bones, especially postcranial ones. The result is that many postcranial bones are crushed and distorted. Whereas length and width measurements tend to be affected less by the plastic distortion, thickness measurements of limb bones can be highly unreliable. For example, the 705 mm length of one femur (TMP 1989.55.1507) is probably not too different from its length when the animal was alive. The 104 mm transverse shaft width is also probably close to its true measurement, whereas the anteroposterior shaft diameter of 37 mm indicates a considerable degree of vertical flattening as the sediments dewatered. This level contains occasional clay balls and may represent a crevasse splay. The bone bed matrix contains many comminuted plant fragments, compressed carbonized wood branches and logs, and seeds. Rare deciduous leaves (occasionally intact) occur just below the bone bed. Pieces of hard and clear fossil resin, some greater than 5 mm in diameter, are abundant throughout the bone layer. The amber has not been studied fully, but several pieces have yielded fossil insects. Some downward-branching carbonized root systems in hard ripple-marked siltstone (of the same lithology as the bone layer) are present in situ just above the bone bed. Silicified wood is absent from the section. The site is in a heavily forested area, where water seepage causes soil creep and slumping. Fossils in many areas of the bone bed have developed fractures that have been subsequently filled in by mud. This situation, combined with frost heave, sometimes separates the fragmented parts of some bones.

Taphonomy

Accurate diagrams showing relative positions of most of the bones in place have been prepared (Fig. 3, Ralrick and Tanke, this volume).

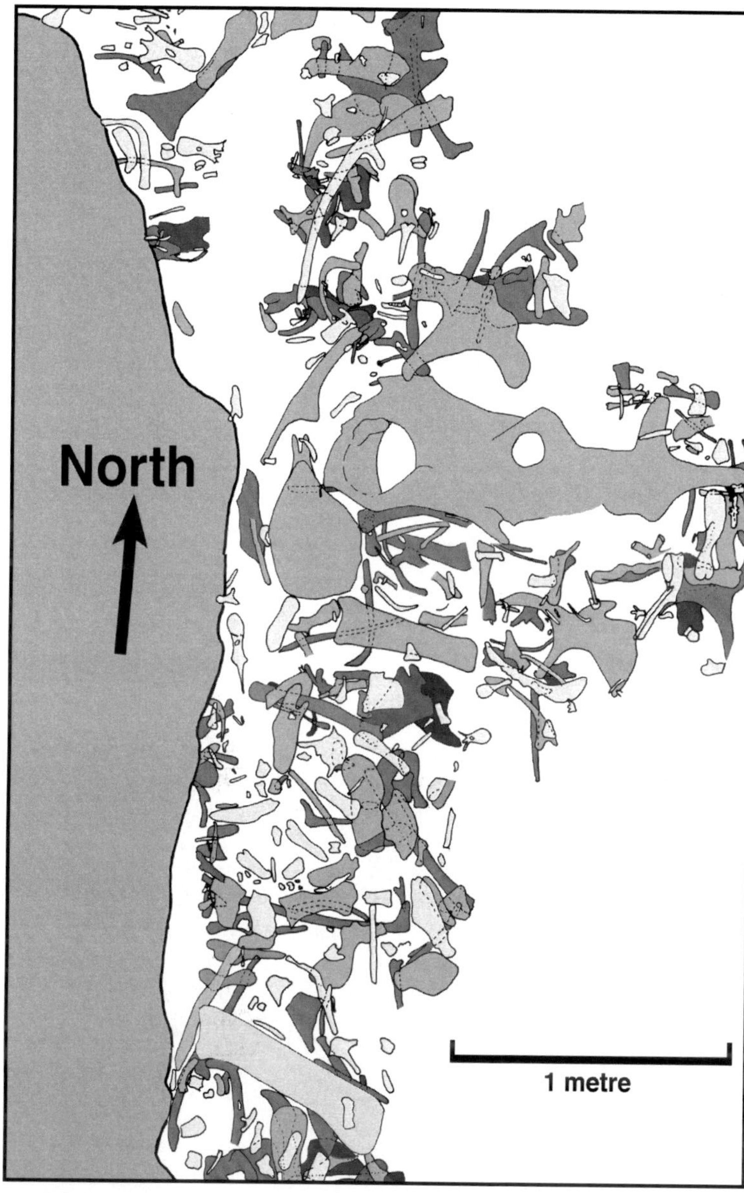

Fig. 3. Quarry map of a portion of the bone bed on Pipestone Creek that was worked in 1986. The eroded cliff face and Pipestone Creek is to the left, and the bone bed is covered by overburden to the right. The skull visible in the drawing is TMP 1986.55.258 (the holotype). Lightly shaded bones generally overlap those that are more heavily shaded.

The bones are stacked in closely packed profusion with close to 200 bones/m^2, ranging from small teeth and bone fragments to massive skulls, in the thickest part of the bone bed. On average, the major part of a skull is found in every 1.5–2 m^2. Bone density is so high that, in places, the bone layer contains less than 20% sediment by volume. Disarticulation of elements is almost universal. Most immature specimens, small fragments, and teeth have been found near the top of the accumulation. All identifiable waterworn bone fragments are ceratopsian. Skulls always lie on top of other bones; because of their great bulk, they usually protrude into the hard overlying siltstone. Most bones lie with their broadest surfaces parallel to the bedding plane and have been compressed vertically, often without visible breakage. Even the massive skulls are somewhat flattened. Waterworn chips of bone and short segments of ossified tendons are encountered regularly, but most bones are essentially complete. Similar to the conditions at Scabby Butte (Langston 1975), diagenetic adjustment in the silty matrix has polished the high points of some of the larger bones. However, much of the bone retains its fine surface texture. Some bones fragmented in situ exhibit microfaulting with slickenside surfaces. Edges and ends of many elements are broken and polished. Marks on some surfaces could have been made by either sediment compaction or preburial transport. Parallel microscratches on most bones indicate some postmortem trampling, but very few of the bones show predator or scavenger tooth marks. Some bones, especially the skulls, bear concretionary coatings that separate from the bone with difficulty. Many

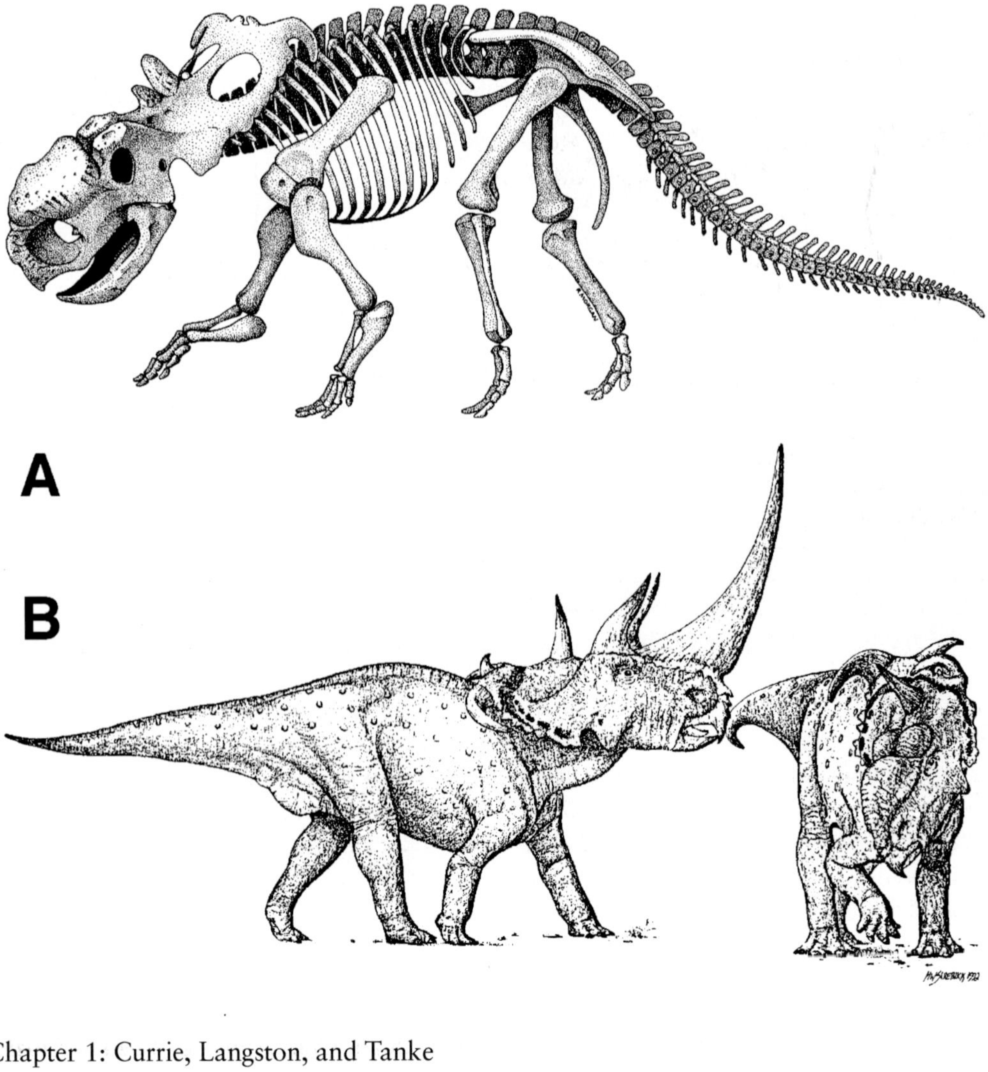

Fig. 4. (A) Skeletal reconstruction of Pachyrhinosaurus lakustai. *(B) Life reconstruction showing two possible options for keratinous coverings of the nasal and suborbital bosses. Note that Langston does not subscribe to the idea of keratinous horns. Animal on the right also displays a pathological frill, based on TMP 1987.55.210.*

pieces of frills were distributed randomly throughout the bone bed, and no complete frills have been found directly associated with skulls. In spite of the disarticulation, almost the entire skeleton is represented, which makes it easy to reconstruct the animal (Fig. 4).

Virtually every bone of the pachyrhinosaur skeleton is represented in the Pipestone Creek collection. Only subadult maxillae and chevrons of all sizes were noticeably rare. This suggests that the specimens were buried close to where the carcasses decomposed and disarticulated. The specimens represent individuals ranging in length from 1.5 m to adults possibly exceeding 6 m in length.

In addition to the pervasive pachyrhinosaur remains, other vertebrates represented in the bone bed include a few holostean bones and scales, a chelydrid turtle neural plate (TMP 1989.55.1514), a varanid (cf. *Paleosaniwa*) vertebra (TMP 1989.55.330), a crocodilian tooth (TMP 1987.55.339) and two osteoscutes (TMP 1987.55.319, Fig. 5; TMP 1989.55.1604), a dromaeosaurid claw (TMP 1986.55.188) and caudal vertebra (TMP 1988.55.129), an elongate *Saurornitholestes* frontal (TMP 1989.55.129), and several dozen small theropod (*Troodon* and *Saurornitholestes*) and tyrannosaurid (cf. Albertosaurinae) teeth.

Fig. 5. Crocodile osteoderm (TMP 1987.55.319) recovered from the Pipestone Creek bone bed.

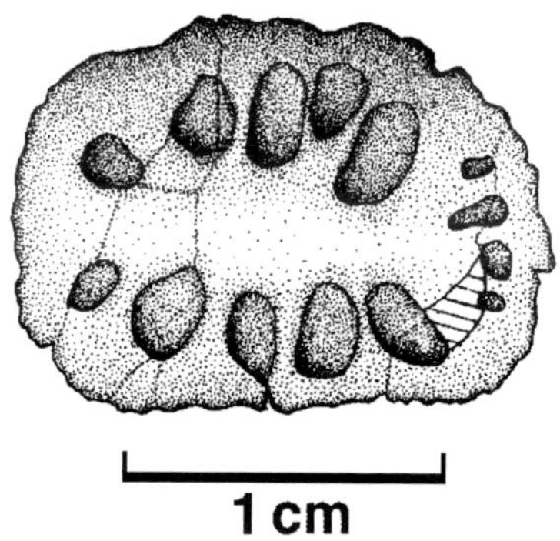

Institutional Abbreviations: AMNH, American Museum of Natural History, New York, USA; CMN, Canadian Museum of Nature, Ottawa, Canada; MOR, Museum of the Rockies, Bozeman, Montana, USA; ROM, Royal Ontario Museum, Toronto, Canada; TMP, Royal Tyrrell Museum of Palaeontology, Drumheller, Canada; UALVP, University of Alberta Laboratory of Vertebrate Palaeontology, Edmonton, Canada.

An unnumbered skull of *P. canadensis* (Langston 1967, 1968) in the Drumheller Valley Interpretive Centre (formerly the Drumheller and District Museum) in the town of Drumheller (Alberta) is referred to as "the Drumheller skull."

Materials and Methods

About 35% of the material collected has been prepared for study and display. Most of the 15 skulls (TMP 1985.112.1, TMP 1986.55.111, TMP 1986.55.206, TMP 1986.55.258, TMP 1987.55.156, TMP 1987.55.196, TMP 1987.55.199, TMP 1987.55.285, TMP 1987.55.304, TMP 1987.55.320, TMP 1989.55.188, TMP 1989.55.427, TMP 1989.55.918, TMP 1989.55.1234, TMP 2002.29.1) were prepared for this study. None of the skulls is complete (for the purposes of this study, a skull is considered to consist of at least the fused nasal boss). In addition to these, the number of right nasals (TMP 1985.112.110, TMP 1989.55.53, TMP 1989.55.191, TMP 1989.55.728, TMP 1989.55.958, TMP 1989.55.1240) of immature animals increases the minimum number of skulls represented in the bone bed to 21. The remainder of the prepared material includes a substantial number of individual cranial bones from the juvenile and subadult skulls, incomplete parietal frills, and numerous postcranial elements.

Specimens collected by Tyrrell Museum field parties bear accession Nos. TMP 1983.55, TMP 1985.112 (Lakusta collection), TMP 1986.55, TMP 1987.55, TMP 1988.55, TMP 1989.55, and TMP 2002.29. The first set of digits in each accession number represents the year that the specimens were collected. It is followed by a batch number that refers to the accessions from a single locality within any given year. Specimens collected in each year are numbered consecutively.

Our descriptions are based on the prepared specimens from the bone bed, with emphasis on the holotype skull of *Pachyrhinosaurus lakustai* (TMP 1986.55.258). Owing to distortion, incomplete preparation, and missing edges in some specimens, measurements given are often approximate but, unless indicated, are not estimated.

Most sutures between dermal elements in mature *P. lakustai* skulls are usually difficult to discern, externally at least. Identification of bone boundaries (Fig. 6) is based primarily on two adult skulls (TMP 1987.55.156, TMP 1989.55.188) in which some sutures were evident, but others were obliterated by fusion. Individual dissociated bones of younger animals also provide much information on the interrelationships of various elements.

Owing to the distinctive morphology of the pachyrhinosaur skull, emphasis is put on the unusual hypertrophied structural units—the nasal boss, supraorbital boss, and the squamosal–parietal frill.

Size range in the Pipestone assemblage

An ontogenetic series can be identified in the Pipestone Creek assemblage (Tanke 1988; Sampson and Tanke 1990; Sampson et al. 1997). Based on the lengths of the humeri (200–555 mm), femora (255–790 mm), and tibiae (208–550 mm), the biggest individuals represented in the bone bed were linearly between 2 and 3 times larger than the smallest. Some cranial elements reveal that there were individuals smaller than this range. The smallest individuals have bones that are approximately 20% smaller than the equivalents in the holotype of *Brachyceratops montanensis* (Gilmore 1917) but are larger than juveniles represented by partial skulls of *Centrosaurus* (Dodson and Currie 1988) and *Triceratops* (Goodwin et al. 2006; Horner and Goodwin 2006). The largest individuals are the size of adult *Centrosaurus* (Tanke 1988).

Large specimens with integrated nasofrontal and supraorbital bosses are viewed as adult individuals, whereas disarticulated cranial bones are regarded as subadult or juvenile, depending on size, surface bone texture, and cranial ornamentation. The smallest animals of about 1.5 m in total body length are referred to as juveniles. Subadults are larger than 1.5 m but still retain separate nasals, juvenile (or a

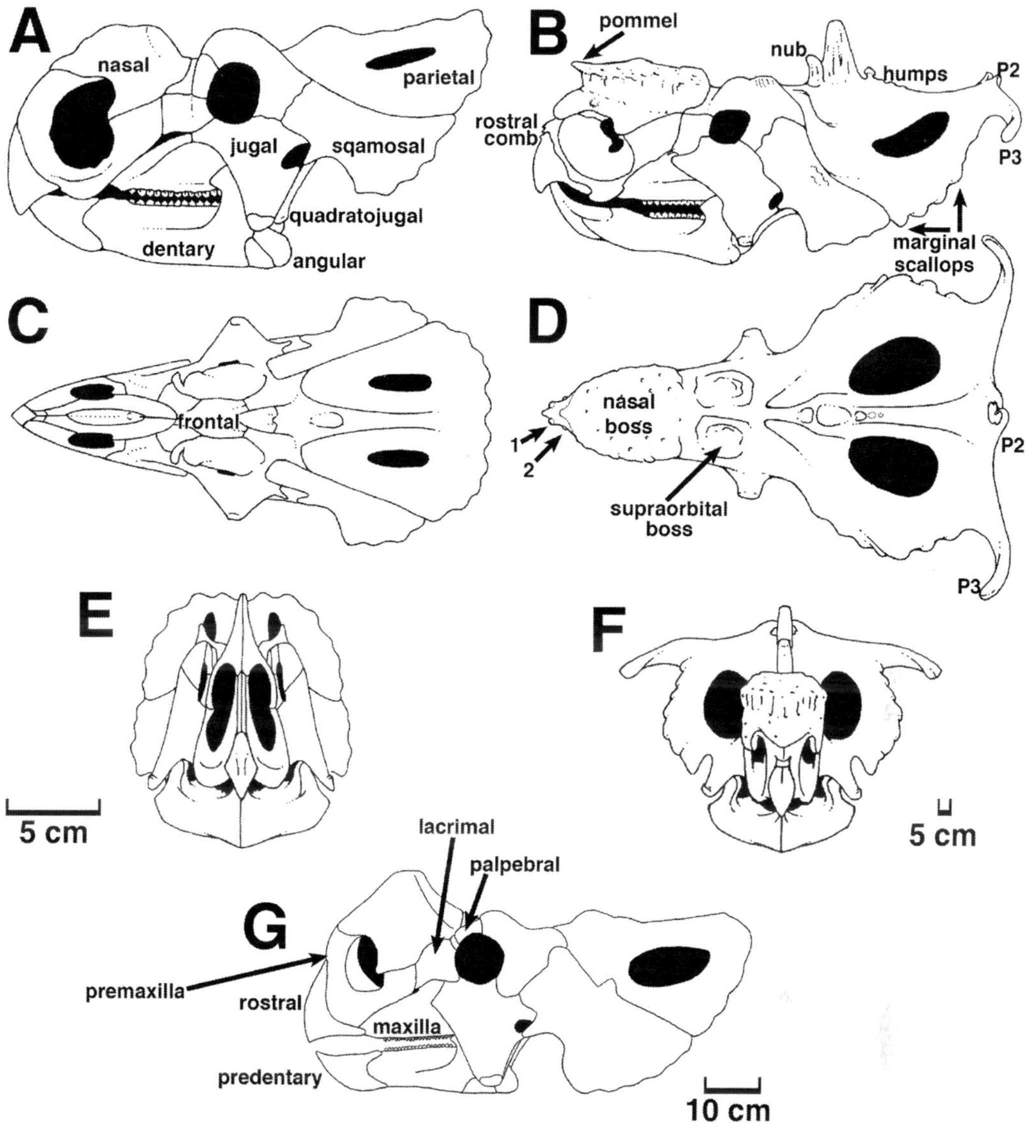

mosaic of juvenile and adult) bone texture, unadorned parietals, and unfused neural arches. All large skulls (greater than 80 cm from rostrum to occipital condyle) are approximately the same size (Table 2) and presumably correlate with the largest limb bones.

Fig. 6. Reconstructed juvenile (A, C, and E) and adult (B, D, and F) skulls of Pachyrhinosaurus lakustai *in lateral (A and B), dorsal (C and D), and anterior (E and F) views. (G) Subadult skull in lateral view. The positions in* Pachyrhinosaurus canadensis *of the supranasal boss (1) and fist-sized knob (2) of Sternberg (1950) are shown. P2 and P3 are the second and third processes defined by Sampson (1995).*

Table 2. Measurements (mm) of pachyrhinosaur skulls following system used by Langston (1975, table 1).

	A	B	C	D	E	F	G	H	I
CMN 8866	570	145	?	?	?	?	?	?	?
CMN 8867	560	160	?	?	115e	?	310	135	170
CMN 9485	530	144	560	295	95e	430	302	147	180
UCalifBerkeley V73078/117105	?	?	?	?	94	?	?	?	?
Drumheller skull	618	172	680	340	100	440	300	136	154
TMP 1982.51.01	?	?	?	?	?	?	?	?	?
TMP 1985.112.1	?	?	?	?	?	?	?	115e	?
TMP 1986.55.111	?	?	?	?	38	325	225	140	190e
TMP 1986.55.206	?	?	?	?	?	?	?	?	?
TMP 1986.55.258	420	140	450	255	65	355	270	160e	190
TMP 1987.55.156	445	160	470+	271	80	315	245	137	?
TMP 1987.55.196	?	?	?	?	?	?	?	?	?
TMP 1987.55.199	?	160	480	252	?	370	?	?	?
TMP 1987.55.285	400	135	445	205	?	380	280	?	?
TMP 1987.55.304	?	98+	?	145+	98	372	280	135	?
TMP 1987.55.320	?	130	?	195	80e	340	255	135	?
TMP 1989.55.188	383	130	410	233	70	318	244	147	?
TMP 1989.55.427	?	?	?	?	?	?	?	?	?
TMP 1989.55.918	?	?	?	?	?	?	?	?	?
TMP 1989.55.1234	470	145	600	260	75	350	245	119	170e

Note: The first five skulls in the list are *Pachyrhinosaurus canadensis*, UCalifBerkeley is a *Pachyrhinosaurus* skull from Alaska that is currently at the University of California (Berkeley), and the rest are *Pachyrhinosaurus lakustai*. The last measurement ("T", length of nasal boss) is new. Abbreviations: A, top of nasal boss to ventral limit of maxilla; B, vertical height of narial opening as seen in lateral view; C, maximum vertical height of skull; cr, crushed; D, vertical height of nasal fossa; E, top of orbit to tip of supraorbital boss; e, estimate; F, rostrum to posterior edge of naris; G, posterior edge of naris to anterior margin of orbit; H, anteroposterior length of orbit; I, posterior edge of orbit to supratemporal fenestra; J, antorbital length; K, height of orbit; L, posterior edge of naris to posterior edge of narial boss; M, bottom of orbit to bottom of jugal; N, greatest width of expansions behind nares; O, width across lowest flanges of premaxillae; P, least transverse width of narial bridge; Q, greatest width of narial aperture; R, greatest width of nasal boss; S, least transverse width of face between nares and orbits; T, length of nasal boss; +, measurement is too low because of crushing.

The largest individuals were presumably adult because of cranial boss architecture and the remarkable frill ornamentation (Tanke 1988) and because cranial bones have mature surface texture (Sampson et al. 1997). The large skulls remained relatively intact owing either to fusion or to sutural complexity not yet developed in smaller skulls, whose elements are always disarticulated. The smallest examples (Figs. 6A, 6C, 6E) and those of the next larger group (Fig. 6G) exhibit none of the remarkable horn and frill specializations seen in the largest individuals (Figs. 6B, 6D, 6F). The subadults (Fig. 6G) show various precursors of the remarkable cranial features of the adults. The cranial bosses and the bony spikes and horns on the parietals appear suddenly

J	K	L	M	N	O	P	Q	R	S	T
?	?	?	?	360e	?	?	?	410	?	?
?	140	400	335	?	?	?	?	?	?	?
732	146	420	?	310	133	84	195	390	360	540
?	87	?	?	?	?	210	?	210	112	380
876	180	340	?	265	140	70	190	350	330	610
?	?	?	?	?	?	?	?	350	?	?
?	?	286	?	?	?	?	70e	309	210e	440
580	125	285	?	?	?	47	?	240+	?	458
?	?	?	?	220	?	?	?	317	230	467
625	120	375	?	170cr	?	51	?	170	?	500
570	138	270	?	154	108	67	108	240	110+	485
?	?	?	?	?	?	?	?	400e	?	420e
?	?	?	?	?	?	57	97	150+	?	?
650	?	280	?	280	130e	70	?	300e	?	494
645	?	340	?	240	?	92	150e	285	?	490
600	130	355	?	?	?	30	80cr	?	?	465+
580	126	302	283	126+	105+	45	77+	156+	95+	475
?	?	?	?	?	?	?	?	315	?	420+
?	?	?	?	?	?	75	140e	250	?	470
550	125	340	?	136+	85+	60	77+	170+	115+	490

in the largest (presumably adult) individuals, in what constitutes an osteological transformation unequaled in other dinosaurs. The largest remains from Pipestone Creek are between 15% and 21% smaller than examples of *P. canadensis* from southern Alberta. Juvenile cranial features are remarkably similar to those of *Brachyceratops montanensis*, an immature centrosaurine, and "*Monoclonius*" (Sampson et al. 1997). This demonstrates that juvenile centrosaurines shared a common skull morph and that specific characters appeared only at sexual maturity (Tanke 1988; Sampson and Tanke 1990; Sampson et al. 1997).

Monotypy at Pipestone Creek

Large dinosaur bone beds are common in Upper Cretaceous deposits in North America (Currie 1981; Rogers 1990; Fiorillo 1991; Varricchio and Horner 1993; Ryan 1994; Ryan et al. 2001; Eberth and Getty 2005; Rogers et al. 2007). In many instances, the remains of a single taxon dominate specific accumulations. Common practice is to assign closely similar material from such deposits to one species, attributing any morphological differences to intraspecific variability. This is justified because multiple closely related species usually do not inhabit the same ecological niches and, therefore, do not end up in the same bone beds, especially if these are the result of rapid burial. This approach seems particularly applicable to concentrations of such large animals as ceratopsians. Indeed, extensive *Anchiceratops* (Sternberg 1926), "*Brachyceratops*" (Gilmore 1917), *Centrosaurus* (Currie 1981; Currie and Dodson 1984; Eberth et al. 1990; Ryan et al. 2001), *Chasmosaurus* (Lehman 1989, 1990), *Einiosaurus* (Rogers 1990; Sampson 1995, 1997*a*, 1997*b*), *Pachyrhinosaurus* (Langston 1975, 1976; May and Gangloff 1999; Fiorillo and Gangloff 2003; Fiorillo 2004), *Styracosaurus* (Visser 1986; Wood et al. 1988; Andersen 2000), "*Torosaurus*" (Gilmore 1946; Farke and Williamson 2005), and *Zuniceratops* (Wolfe and Kirkland 1998; Wolfe 2000; Wolfe et al. 2004) bone beds are known. On the other hand, intrusive elements, such as an *Anchiceratops* in a *Pachyrhinosaurus*–hadrosaur bone bed in southern Alberta (Langston 1975), sometimes are found in predominantly monotypic accumulations. Thus, although the extremes in morphology (especially in the nasal and postorbital horncores or bosses) are radically different in the Pipestone Creek ceratopsian assemblage, all ceratopsian remains from this site are assumed (as a working hypothesis) to be derived from a single pachyrhinosaur population. Morphological deviations are considered to be attributable to ontogenetic factors, individual variation, and possibly sexual dimorphism. Analysis of specimens in the assemblage defines an ontogenetic sequence from juvenile to adult, and highly variable frill morphology suggests the possibility of distinct male and female morphs.

Systematic palaeontology

Dinosauria Owen, 1842
Ornithischia Seeley, 1888
Ceratopsia Marsh, 1890
Neoceratopsia Sereno, 1986
Ceratopsoidea Hay, 1902
Ceratopsidae Marsh, 1888
Centrosaurinae Lambe, 1915 "Pachyrhinosaurs"
(*Pachyrhinosaurus* + *Achelousaurus* clade)

Amended Diagnosis: Pachyrhinosaurs are a distinct clade within the Centrosaurinae that were once considered as a distinct ceratopsid subfamily, the Pachyrhinosaurinae, by von Huene (1950). Immature individuals have elongate, low, recurved, and longitudinally divided nasal horn cores. Juveniles and subadults have small oval supraorbital horn cores. Adult skulls have dorsolaterally expanded and thickened nasal bones constituting a massive nasal boss whose elevated margins may contain relict "juvenile" horn cores dorsolaterally. Smaller, flattened bosses composed of postorbital and frontal bones replace the juvenile supraorbital horn cores in adults. A transverse tunnel created by junction (but not fusion) of postorbital and nasal bosses occurs at least in large individuals represented by the Drumheller skull and other specimens of *P. canadensis* (Langston 1975). Other facial ornamentations, some possibly created by osteoderms, are variably developed in larger individuals. The parietal–squamosal frill is short with a highly variable array of marginal cornua on the parietals in larger skulls. The median parietal bar is variously ornamented by a series of sagittal humps or by straight or curved, deeply fluted horn cores. In mature specimens, strongly expressed "scallops" are present on the lateral edges of the parietals and squamosals, with or without epoccipital ossifications (dependent on fusion). The pachyrhinosaur squamosal has four to six marginal scallops. The first, most medial parietal process (P1 of Sampson et al. 1997) of *Centrosaurus* and *Styracosaurus* is not present in pachyrhinosaurs. An epijugal bone may fuse to the jugal. Large external nares contain the thickest premaxillary narial septum known in ceratopsians. The lateral facial plates of the nasals expand transversely, imparting an unusual flared appearance to the facial region when viewed from the front. A long medially curved finger-like process of the nasal bone projects into the external naris posteromedially. A medially directed flange of the premaxilla partially divides the large void within the facial region, perhaps separating the narial passage below from a larger cavum nasi proprium above. The antorbital fenestra is present but small. With the exception of the jugal, all facial bones are greatly thickened. A massive palpebral element is integrated into the dorsal rim of the orbit. No "antorbital buttress" is present around the anterior margin of the orbit. The squamosal bone is relatively shorter than in other Ceratopsidae, and the "otic notch" is correspondingly abbreviated. The quadratic canal is largely confined to the quadratojugal. The braincase walls are greatly thickened compared with that of other ceratopsids. The exits for cranial nerves III, IV, and VI leave the laterosphenoid through separate foramina; the exit for cranial nerve VII is remarkably small. A tiny foramen for the internal carotid artery lies farther ventrally than is usual in ceratopsids. The occipital condyle is oriented posteroventrally. The pachyrhinosaur dentary is deeper, and the coronoid process is more powerfully developed than in other ceratopsids. The postcranial skeleton is massively constructed on the order of *Triceratops* (possibly excepting the humerus and tibia). Neural spines of the coalesced anterior cervical vertebrae are lower than in other large ceratopsids.

Stratigraphic Distribution: Horseshoe Canyon Formation, Prince Creek Formation, St. Mary River Formation, Two Medicine Formation, Wapiti Formation.

Age: Upper Cretaceous (Campanian – lower Maastrichtian).

Geographic Distribution: Southern and central Alberta, Canada; Montana, northern Alaska, USA.

Included Species: Achelousaurus horneri Sampson, 1995; *Pachyrhinosaurus canadensis* Sternberg, 1950; ***Pachyrhinosaurus lakustai*** sp. nov.

Holotype: TMP 1986.55.258. A transversely crushed adult skull, lacking most of the jugals, quadratojugals, quadrates, squamosals, and most of the parietals. Collected by Darren Tanke, 1986.

Topotypic Material: About 2500 bones comprising an ontogenetic series of a single population.

Locality and Horizon: East bank of Pipestone Creek (10 m above water level), 19 km southwest of Grande Prairie, Alberta, Canada. Additional material was recovered 120 m downstream on the west bank of this creek. Wapiti Formation, in Upper Campanian, soft, dark gray carbonaceous siltstone. Late Cretaceous, Late Campanian age.

Etymology: "lakustai" after Al Lakusta, who discovered and reported the bone bed.

Diagnosis: Smaller than *P. canadensis*; adults with relatively smaller rostral bone; "rostral comb" on rostral and premaxillary bones; single transverse shelf of bone on the medial side of the premaxilla compared with a pair of shelves in all other centrosaurines; nasal boss and other anterior facial ornamentations (including "fist-sized knobs") relatively less hypertrophied than in *P. canadensis*; short robust premaxillary process projects straight forward into nares, whereas it is a long slender, finger-like process in *P. canadensis* that extends forward and then curves 90° medially; nasal boss texture is strongly and regularly developed fluting on sides of adult nasal bosses; fluting is angled anterodorsally at 45° in *P. lakustai* but is irregularly developed and vertical in *P. canadensis*; anterior end of the nasal boss projects freely in a spout-like fashion above the narial bridge; orbital and nasal bosses are separated by a smooth T-shaped groove; grooves at bases of horns curving together on midline of parietal (marginal process 2 of Sampson 1995; Sampson et al. 1997); parietal bar curves slightly upwards, whereas the parietal bar in *P. canadensis* has an essentially flat profile in lateral view; one to three median horns are present near the base of midparietal bar in subadult and adult stages; lateral parietal spike twists anteroventrally instead of dorsolaterally.

Description

Sampson (1995) observes that, from a systematic point of view, diagnostic character states in Centrosaurinae are limited to the skull roof. Indeed, it has been long apparent that, within this subfamily, individual cranial elements without ornamentations are virtually indistinguishable. However, *Pachyrhinosaurus* proves the exception to this generalization, because relative proportions and various hypertrophies of facial elements are distinctive of this genus. The following notes address some such features seen in nonornamented skull bones and in additional states in some elements described in connection with the various cranial bosses and the frill. Isolated elements provide interesting details about the paths of sutures—usually obliterated in intact skulls—and the nature of these sutures.

The holotype skull (Figs. 7A, 7B) is missing most of the cheek region, the squamosals, and the lateral and transverse rami of the

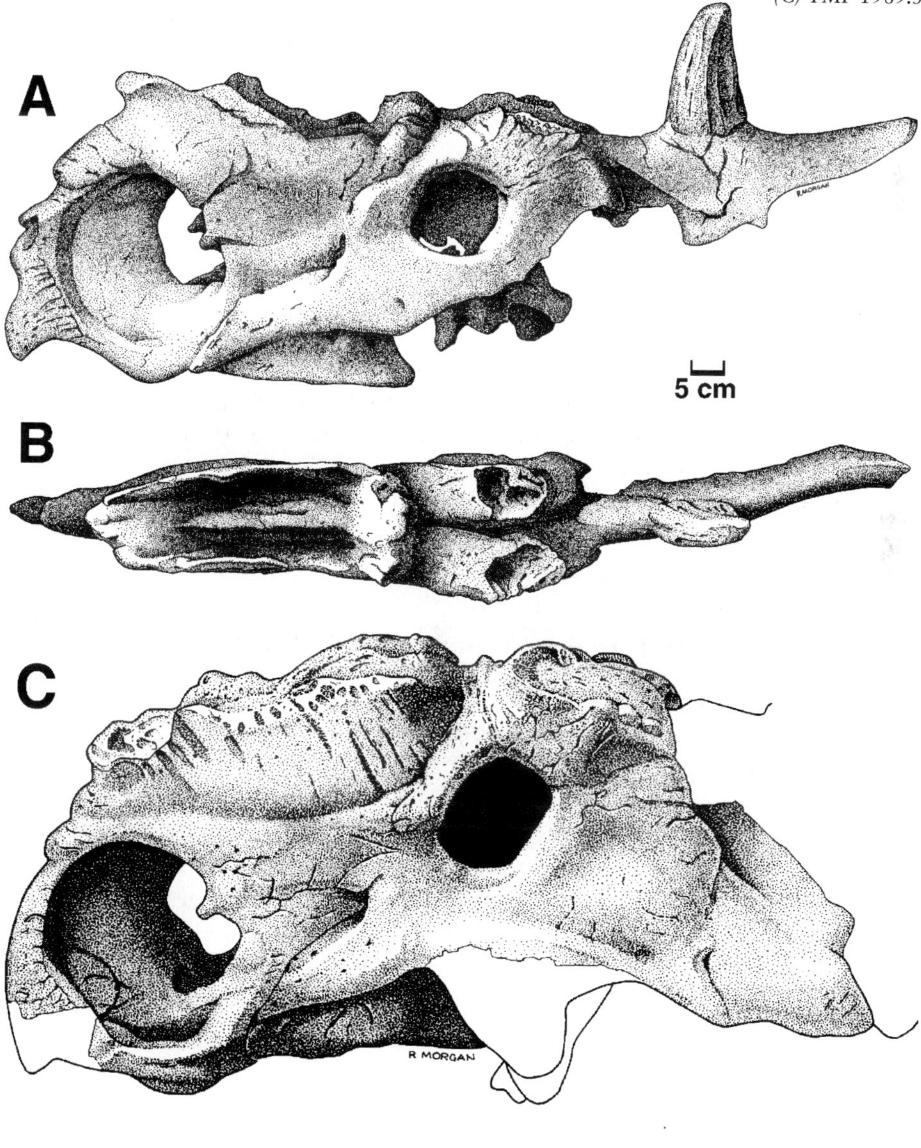

Fig. 7. Pachyrhinosaurus lakustai *holotype skull (TMP 1986.55.258) in left lateral (A) and dorsal (B) views. The dorsal view shows the extreme transverse compression. (C) TMP 1989.55.1234.*

Fig. 8. Pachyrhinosaurus lakustai *skulls in left lateral view. (A) TMP 1987.55.285; (B) TMP 1989.55.188; (C) TMP 1987.55.320; (D) TMP 1987.55.156; (E) TMP 1986.55.111; (F) composite skull that is composed mostly of TMP 1989.55.1234 but includes TMP 1989.55.260 (posterior end of left squamosal), TMP 1986.55.164 (right lateral parietal bar), and TMP 1989.55.1241 (back of frill).*

parietals. It is crushed transversely to perhaps about one-half of its natural width. Importantly, it is the only prepared skull that includes the midparietal bar and shows the parietal bar horn in natural position. As preserved, the skull is about 135 cm long from the rostrum to beyond the midlength of the median parietal bar, and has a basal skull length (rostrum to back of occipital condyle) of 87 cm. It is the largest skull from Pipestone Creek so far prepared. It is about 15% smaller than known skulls of *P. canadensis* (Table 2). Another incomplete skull (TMP 1989.55.1234, Fig. 8F), which lacks most of the parieto-squamosal frill and jugal but otherwise is better preserved than TMP 1986.55.258, is about 6% smaller than the holotype.

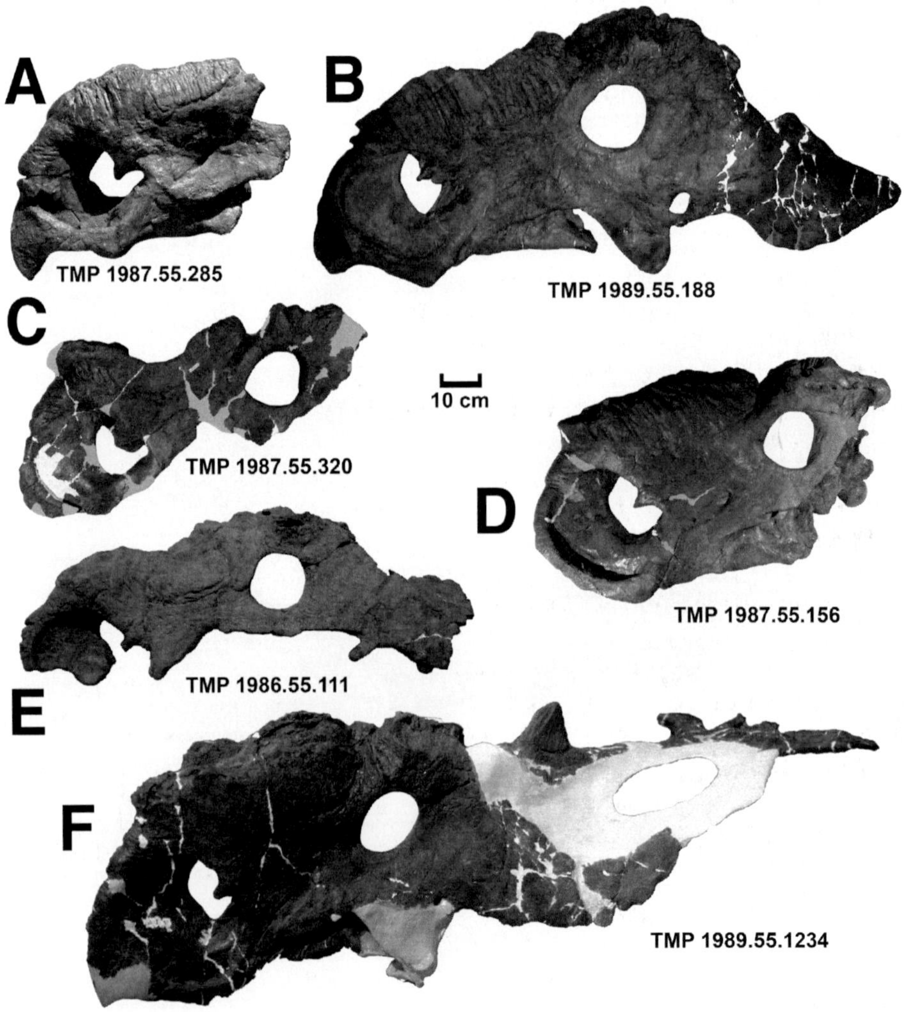

Ceratopsian cranial elements fuse into larger units at different ontogenetic stages. Although these units may remain independent of each other even in the largest individuals, the sutures of the individual bones incorporated into any of the units are often completely obliterated by fusion. Consequently for many elements, descriptions of individual bones are only possible in juveniles and subadults. In the following sections, individual bones are grouped into units (snout, circumorbital, parietosquamosal frill, palatoquadrate, braincase, and mandible) and are described separately. The co-ossified nasal and supraorbital bosses are described separately within their units.

Rostral

The rostral bone (Fig. 9) is variably attached to the premaxillae, and the two bones form the peculiar rostral comb in adults (described after the premaxilla). Two detached rostrals from adults (TMP 1986.55.191, TMP 1987.55.50, Fig. 10) have well-defined, smoothly finished contact surfaces for the premaxillae. Each is as large as the rostral of the holotype, which appears to be firmly fused to the premaxillae. The rostral is broader than might be inferred from the transversely compressed skulls from Pipestone Creek (Fig. 10E). The lateral profile is variable, and some rostrals (such as that of the holotype) appear more hooked and beak-like than others (Figs. 7–9). The lingual surfaces that opposed the predentary, and were presumably horn covered, are broad and separated in part by a low ovate median tubercle (Fig. 10E). Anteriorly, the external surface is deeply sculptured by short chaotic pits and grooves, which grade posteriorly into a more regular pattern of subparallel anteroposterior anastomosing sulci that are perpendicular to the border with the premaxillae. There are no significant differences from the rostrals of other centrosaurines (Fig. 9). Along with the premaxillae, the rostral forms the distinctive "rostral comb" that will be described subsequent to the description of the premaxilla.

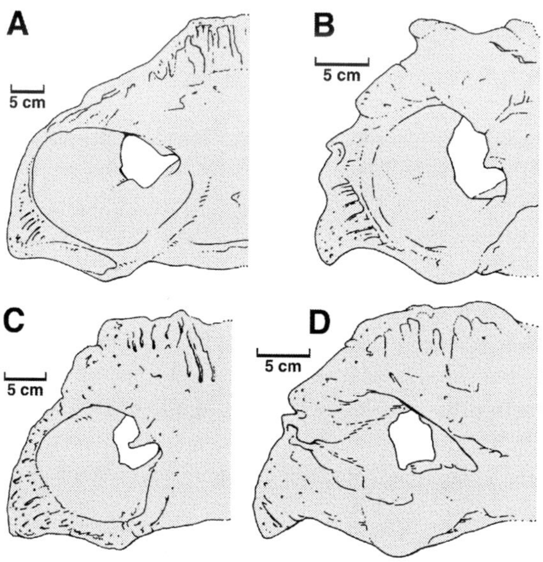

Fig. 9. Pachyrhinosaur rostra illustrating variation in development of the rostral comb and other narial features. (A) Drumheller skull of Pachyrhinosaurus canadensis; *(B) TMP 1986.55.258,* Pachyrhinosaurus lakustai; *(C) NMC 9485,* Pachyrhinosaurus canadensis; *(D) TMP 1987.55.285,* Pachyrhinosaurus lakustai.

Fig. 10. Pachyrhinosaurus lakustai *rostra: TMP 1986.55.191 in anterior (A) and lateral (B and C) views; TMP 1987.55.50 in lateral (D) and ventral (E) aspects.*

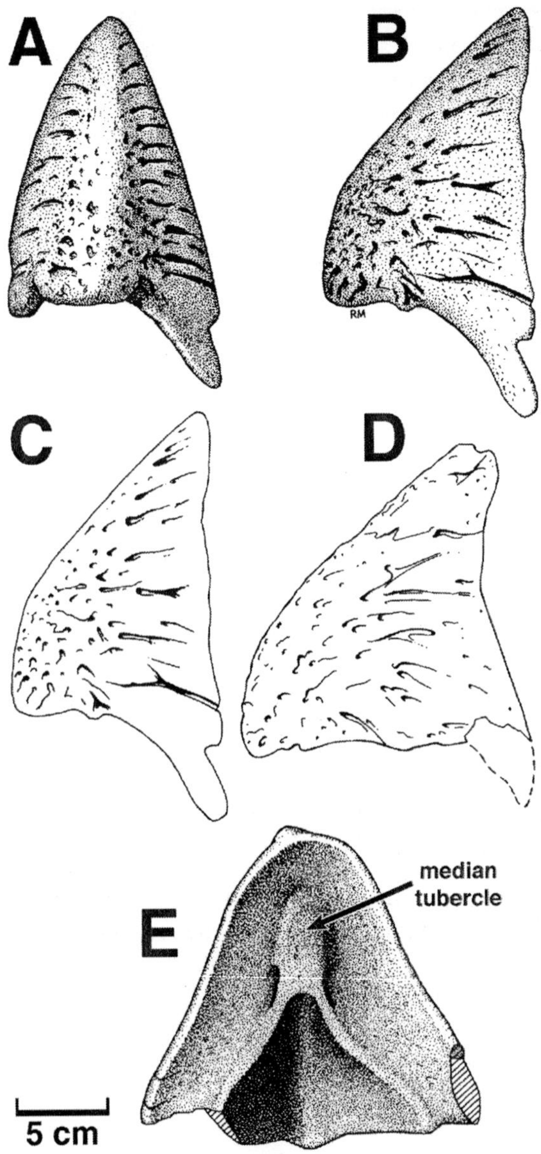

Premaxilla

The premaxilla (Fig. 11) has two distinct segments. One segment contacts the other premaxilla and the nasal to form the narial bridge. The other supports the rostral bone anteriorly, forms the anterior end of the palate ventrally, and houses much of the nasal capsule. The bone is generally similar to that of *P. canadensis* (Sternberg 1950) except for the anterior edge of the narial bridge. The bridge is completely

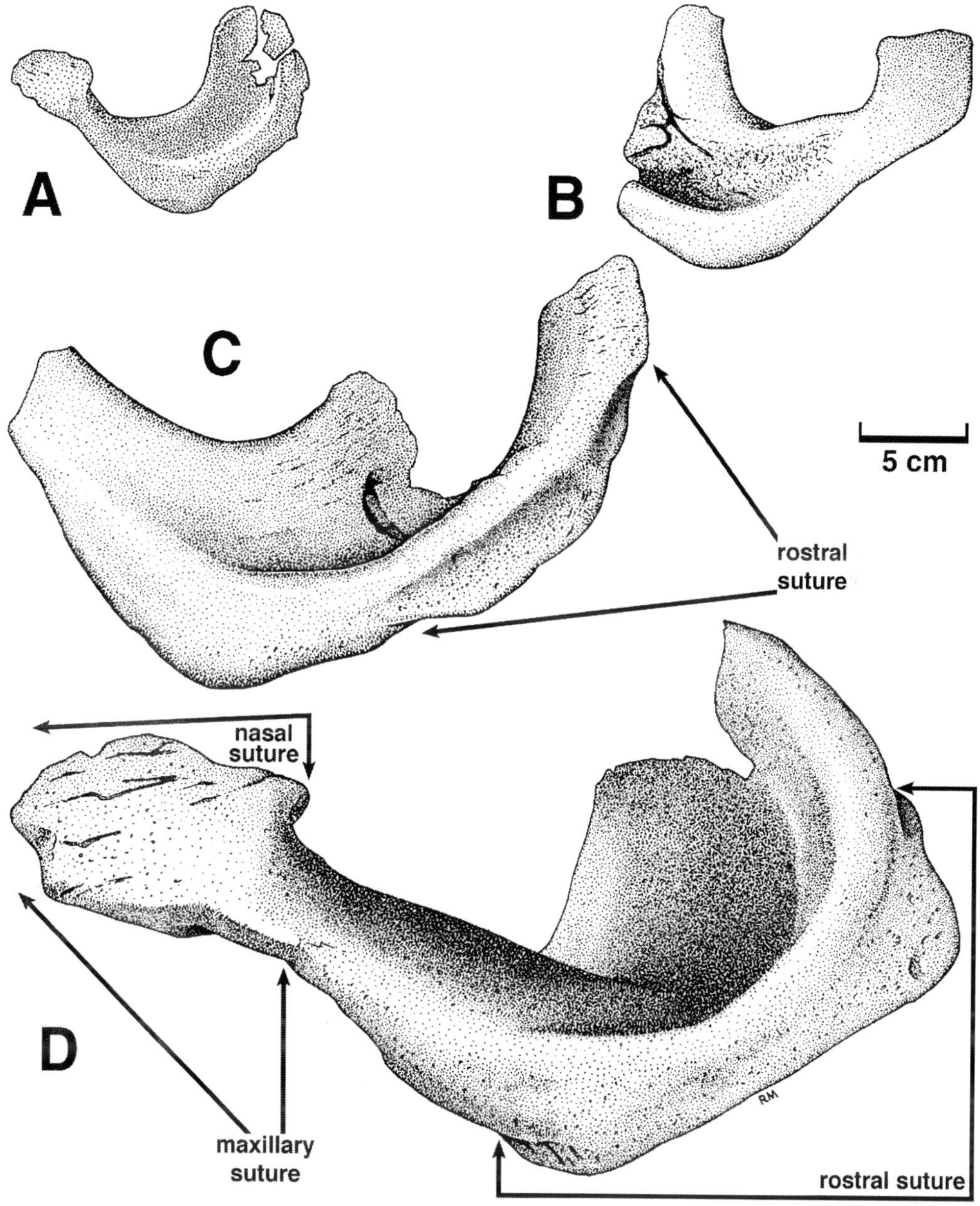

Fig. 11. Pachyrhinosaurus lakustai *premaxillae. (A) Young individual, TMP 1989.55.1199, right lateral view; (B) young individual TMP 1987.55.311, left lateral view; (C) juvenile, TMP 1988.55.233, right lateral view; (D) subadult individual, TMP 1986.55.153, right lateral view (anterior at the right).*

preserved in two skulls (TMP 1986.55.258, holotype, Fig. 7A; TMP 1987.55.285, Fig. 8A) and is partly preserved in two others (TMP 1986.55.111, Fig. 8E; TMP 1987.55.320, Fig. 8C). Additionally, the nasal component of the bridge is preserved in TMP 1985.112.1 and TMP 1986.55.206. Except for the appearance of the rostral comb in some adults and substantial thickening throughout the bone with increasing size, growth of the premaxilla appears to have been isometric. The architecture of the premaxilla provides an important means of distinguishing between chasmosaurine and centrosaurine ceratopsids (Dodson and Currie 1990; Lehman 1990; Dodson et al. 2004). Its elevated, abbreviated, and uncomplicated narial septum places *P. lakustai* and *P. canadensis* clearly within the Centrosaurinae (Langston 1967). On the medial side of the premaxilla is a single modest transverse shelf of bone. This shelf, which is in contact with and is anchored to the anteriorly directed processes of the maxilla, is homologous with a pair of shelves (one anteroventral to the other) that occur in all other centrosaurines (Sampson 1995). There are substantial differences in the direction and inclination of the premaxillary–rostral margin, which can be straighter and less complicated in some examples than in others.

As in other centrosaurines, the posteroventral edge of the premaxillary part of the beak forms a conspicuous ventral angle (Ryan 2003) that extends ventral to the alveolar margin of the maxilla (Figs. 7, 8, 11). The premaxillary septum is relatively thicker and occupies a relatively greater area within the naris than in any other ceratopsian other than *P. canadensis* (Langston 1967). The dorsoposterior flange extends posteriorly to contact the lacrimal as in most other centrosaurines, although occasionally (AMNH 5351, *Centrosaurus nasicornis*), a nasal–maxillary contact prevents the premaxilla from reaching the lacrimal.

Rostral Comb

A unique and variable feature of at least some adult individuals of *P. lakustai* is what we term the rostral comb (Fig. 9). In well-developed examples, the rostral comb comprises two or three massive anteriorly directed hook-like projections that are arranged in tandem on the narial bridge. Such projections do not occur in any other ceratopsian, including *P. canadensis*, where the anterior edge of the nasal is smoothly rounded (Figs. 9A, 9C). The lowest two projections arise from the premaxillae; the third, uppermost projection is formed by the blunt thickened ends of the nasal bones, which clasp the dorsal processes of the fused premaxillae between them. The projections are variable in shape and proportions, appearing asymmetrical on opposite sides of the same specimen. The notches that separate the projections cross the anterior edge of bridge diagonally, reminiscent of the epoccipital scallops on the parietal bone. In each of TMP 1986.55.258 (Figs. 7A, 9B) and TMP 1987.55.285 (Figs. 8A, 9D), a circular excavation with a roughened bottom is present on one side of the notch at the base of the lowest projection. It is conceivable that these excavations are loci for unfused osteoderms. TMP 1987.55.285 (Fig. 9D) has only two projections: one each on the nasal and premaxilla.

The comb-like structure is only seen in some adult skulls, so the ontogenetic stage at which it developed is not known. However, a dissociated and incomplete right premaxilla (TMP 1988.55.233, Fig. 11C) that appears to have been about three-quarters of adult size shows no tendency toward thickening of the bone just above the dorsal end of the rostral bone; thus, like other cranial ornamentations, the comb probably appeared near the onset of adulthood.

Maxilla

The maxilla (Fig. 12) is a complex bone that contacts ectopterygoid, jugal, lacrimal, palatine, premaxilla, pterygoid, and vomer. A strong anteromedial process meets that of the opposite maxilla posterior to where it slots into the premaxilla. As in other centrosaurines, the maxilla is separated from the nasal by the long strap-like posterodorsal ramus of the premaxilla. The medial ascending process is separated from the lateral ascending process by a notch that forms the floor of the relatively small antorbital fenestra. *P. canadensis* supposedly lacks this feature (Sternberg 1950; Langston 1967), but its presence is probably simply obscured by fusion of the flange into the surrounding bone in the skulls from Scabby Butte and Drumheller.

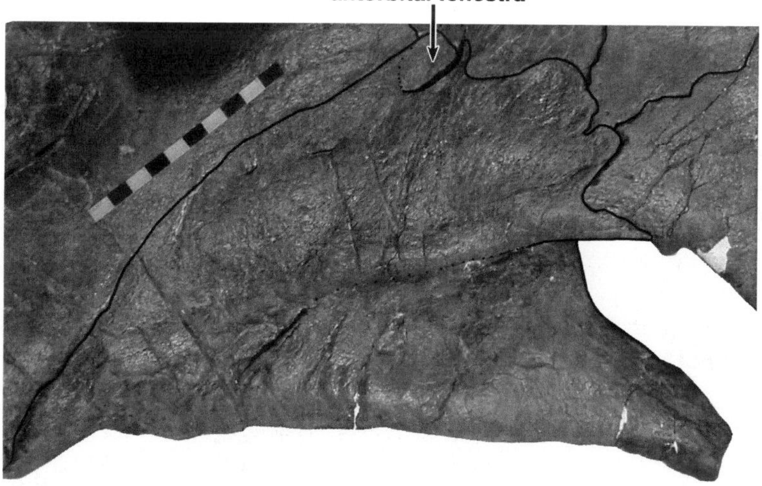

Fig. 12. Pachyrhinosaurus lakustai *left maxilla (TMP 1989.55.188) in lateral view.*

Isolated maxillae range in length from 17 cm (TMP 1989.55.855) to 31 cm (TMP 1989.55.1485), although the best-preserved large maxillae are in articulated skulls (TMP 1989.55.188, Fig. 12). Smaller maxillae have fewer teeth than larger ones (for example, TMP 1989.55.855 is one-half the length of the largest maxillae and has only 17 tooth positions). There are 24 or 25 alveoli preserved in TMP 1989.55.188 and TMP 1989.55.1234, although the total count in both cases may have had a few more tooth positions. These numbers are comparable with other ceratopsids of similar size (25–28 in *Achelousaurus*, 29–35 in *Centrosaurus*, 30 in *Chasmosaurus*, and 24–28 in *Einiosaurus*)

Fig. 13. Pachyrhinosaurus lakustai. *Skull roof with nasal and supraorbital bosses of an adult individual (TMP 1985.112.1) in left lateral (A) and dorsal aspects (B); (C, D, and (E) cross sections in outline. Nasal boss of TMP 1986.55.206 in dorsal (F) and left lateral (I) views with two cross sections (G and H).*

but are lower than in the adults of larger ceratopsids, such as *Triceratops*, which can have as many as 40 tooth positions. Clearly the number of tooth positions increased ontogenetically as in protoceratopsians (Maryanska and Osmólska 1975; Dong and Currie 1993), *Einiosaurus* (Sampson 1993), *Zuniceratops* (Wolfe 2000), hadrosaurs (Horner and Currie 1994), and other ornithischians.

Nasal

Numerous isolated nasal bones from juvenile and subadult *P. lakustai* were recovered. In large individuals, the paired nasals fuse and form a massive boss of bone that is characteristic of pachyrhinosaur skulls (Figs. 13,14). The boss seems to be well developed before the nasals co-ossify, and the entire length of the internasal suture can

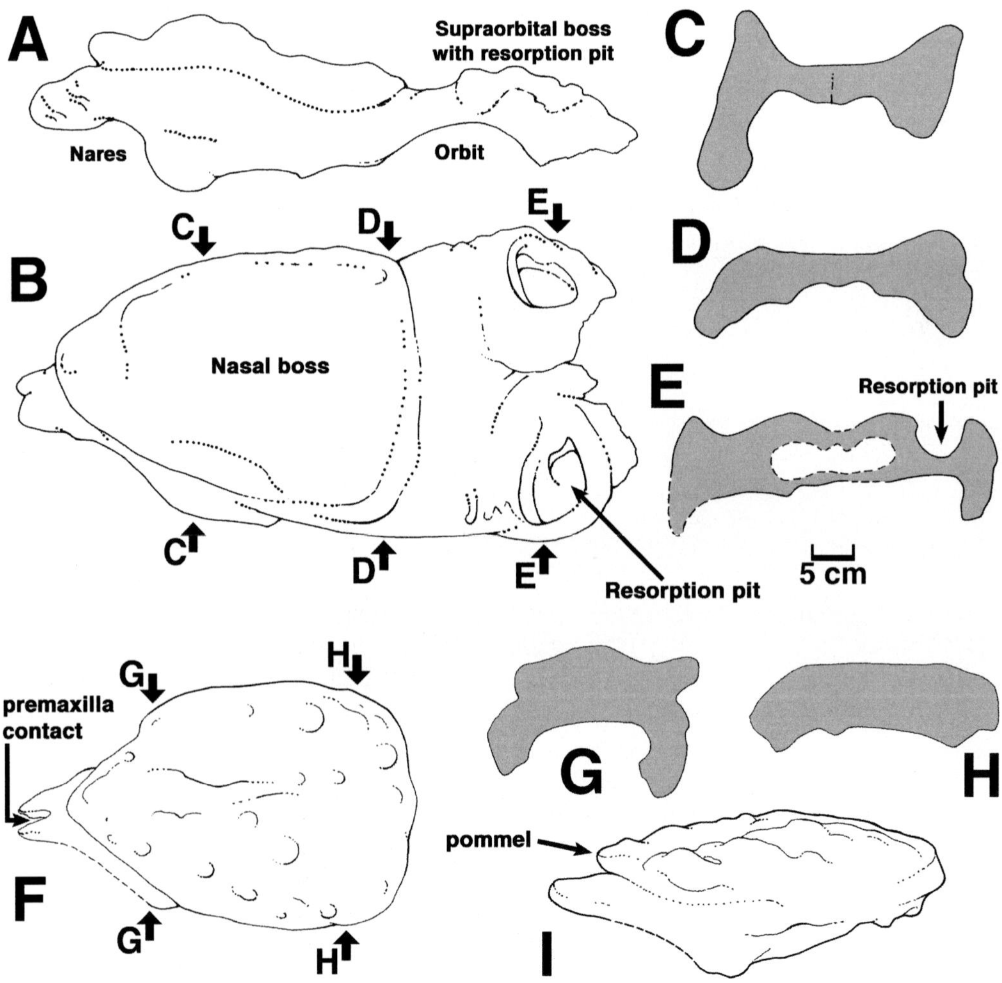

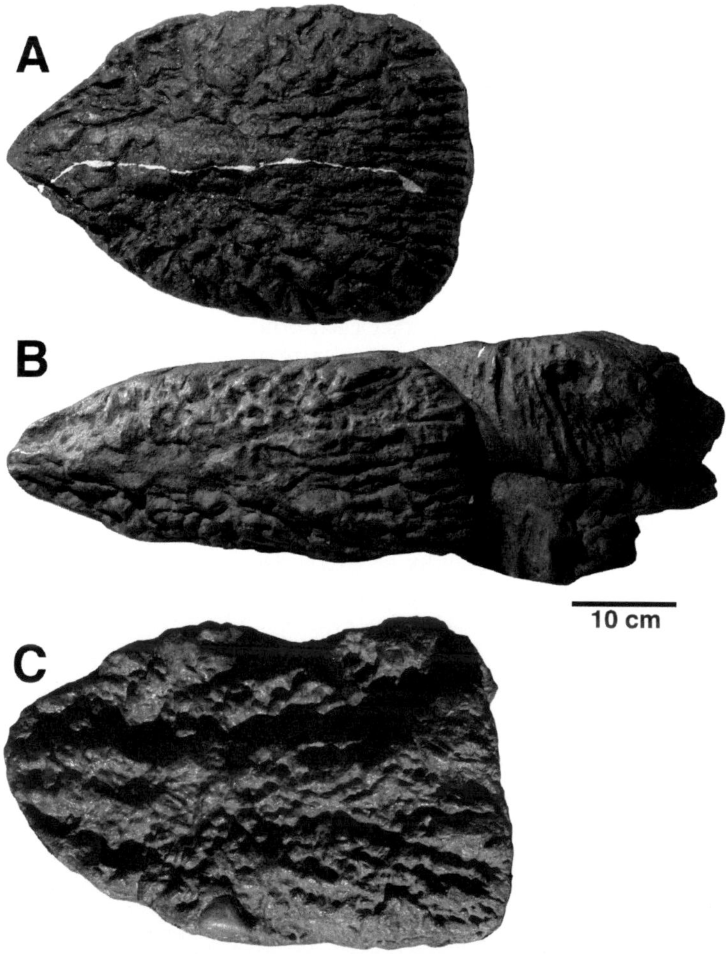

Fig. 14. Pachyrhinosaurus lakustai *nasal bosses in dorsal view: (A) TMP 1986.55.206; (B) TMP 1987.55.156 (crushed lateromedially); (C) TMP 1987.55.285.*

be seen on the ventral surface of one large nasal boss (TMP 1986.55.206). TMP 1989.55.918 is a slightly larger nasal boss, but the suture has closed ventrally for most of the length of the nasals. The massive, coalesced nasal boss extended posteriorly to include parts of the frontals and prefrontals in *P. canadensis* and, therefore, was termed the nasofrontal boss (Langston 1967, 1975). However, it is restricted to the nasals in *Achelousaurus* (Sampson 1995), and no specimens of *P. lakustai* show any indication of prefrontal or frontal involvement in the boss. In TMP 1986.55.206 (Figs. 13F, 13I, 14A) and TMP 1989.55.918, the sutures at the back of the nasal bosses for the prefrontals and frontals are clearly visible. The skulls, disarticulated nasal bosses and supraorbital bosses show that the prefrontals thicken anteriorly and butt into depressions on the posteroventral margins of the nasal bosses. There is some indication that the suture may be partially closed sometimes. However, because the prefrontals and frontals do not take part in the formation of the boss in *P. lakustai*, the structure is referred to as the nasal boss (as opposed to a nasofrontal boss) in this paper. Langston (1975) inferred the possible presence of frontal or prefrontal involvement in the boss of *P. canadensis* by the presence of a low, sagittal ridge at the back of the boss. However, this ridge is also present in *P. lakustai* (TMP 1986.55.206, Fig. 14A), where it coincides with the internasal suture. Therefore, it is conceivable that the nasal boss of *P. canadensis* is also formed only by the nasals. In all probability, the transverse tunnel of Langston (1967) represents the division between the nasal and supraorbital bosses. These units may co-ossify in mature animals, but the nasal and supraorbital bosses only contact each other dorsal to the transverse tunnel. As pointed out by Langston (1967), the universal presence of this feature in well-preserved specimens suggests that some vital soft structure, such as a blood vessel and (or) nerve, was overgrown by the ornamentations.

Nasal bosses are present in all of the adult skulls of *P. lakustai* from Pipestone Creek (Figs. 6, 7, 8), and a number of detached bosses (Figs. 14A, 14C) were also recovered from the bone bed. There is a certain amount of variability in proportions that, in part, can be attributed to plastic deformation of the Pipestone Creek fossils.

Most nasal bosses are distorted, and differences in shape and proportions are difficult to evaluate. On average, bosses from Pipestone Creek skulls may be a little narrower relative to their lengths and to the lengths of the skull roofs than those of *P. canadensis* (Tables 3 and 4). In profile, most *P. lakustai* examples appear somewhat more depressed than *P. canadensis* at midlength. However, these observations are attended by much uncertainty owing to differences in preservation and possible errors in measurement. In plan, adult bosses of *P. lakustai* appear more narrowly triangular than those of *P. canadensis* from Scabby Butte. In lateral view, the dorsal edges of the bosses in many specimens of *P. lakustai* are concave, are higher in front than behind, and have their anterior ends projecting forward and upward (Figs. 7A, 8, 13A–13E). In side view, it is evocative of the pommel of a saddle. Viewed from the front and above the pommel, the front of the boss appears broadly spout-like in some skulls, with raised edges on either side of a sagittal trough. The dorsal rim of the boss is elevated and narrowly horn-like above the posterior ends of the nares (Fig. 13A). However, the dorsal surfaces of these concave bosses invariably appear damaged, and it is highly likely that the concavity is an artifact of preservation. The thin surface bone on *P. lakustai* nasal bosses that covers the highly spongy bone inside was easily damaged after death. The scooped-out nature of some bosses is presumably exaggerated by erosion. Small lumps (<1.5 cm) of trabecular bone that may have been remnants of the damaged regions of some of these bosses were occasionally found in the quarry.

Table 3. Relative proportions of the nasal boss in pachyrhinosaurs.

Specimen	Length of boss/ length of skull roof	Width of boss/ length of skull roof	Width of boss/ length of boss
CMN 9485	0.787	0.528	0.67
UCalifBerkeley	0.76	0.42	0.553
Drumheller	0.754	0.59	0.784
TMP 1985.112.1	0.671	0.471	0.67
TMP 1986.55.111	0.675	—	0.52
TMP 1986.55.206	—	—	0.67
TMP 1986.55.258	—	—	0.34
TMP 1987.55.156	0.70	0.35	0.49
TMP 1987.55.285	—	—	0.61
TMP 1987.55.304	0.69	0.4	0.58
TMP 1989.55.918	—	—	0.53
TMP 1989.55.1234	0.73	—	0.35

Note: The "length of skull roof" is the distance between the front of the nasal boss and the back of the supraorbital bosses.

Table 4. Comparison of the length of the nasal boss with the anteroposterior length of the orbit, the horizontal distance between the back of the external naris and the front of the orbit, and the length of the skull from rostrum to occipital condyle in the three pachyrhinosaur species.

	Length of boss/ length of orbit	Length of boss/ naris–orbit distance	Length of boss/ length of skull
Achelousaurus horneri	2.1	1.3	0.27
P. lakustai	3.6	1.9	0.58
P. canadensis	4.1	1.9	0.61

The palisaded appearance of the sides of the nasal boss is less pronounced than in skulls of *P. canadensis* (Sternberg 1950, pl. 21; Langston 1967, fig. 5). In specimens where the palisades are pronounced, they are inclined anterodorsally in relation to the long axis of the skull. The top of the boss can be more depressed (Figs. 7, 8, 13) and basin-like than in *P. canadensis*. The bone is thinner dorsoventrally, hardly more than 15 mm thick in one adult (TMP 1987.55.320), but approaches *P. canadensis* conditions in another (TMP 1987.55.285, Fig. 14C) where the boss has a greatest thickness of 110 mm. The T-shaped system of ridges believed to define the junction between the nasals, prefrontals, and frontals in *P. canadensis* from Scabby Butte is not present in Pipestone Creek specimens. As in *P. canadensis*, the boss is bounded below by a circumferential groove that is pierced intermittently by small foramina. The sides expand laterally and partly obscure the sides of the face when viewed from above.

A detached boss (TMP 1986.55.206) seems to have suffered relatively little distortion. In plan, this specimen (Figs. 13F–13I, 14A) is more acuminate anteriorly than other nasal bosses. It lacks the "pommel" and spout-like feature anteriorly, although the tip of the boss projects freely for a short distance above the narial bridge. It is possible that dorsoventral crushing may have reduced the height of the boss and accentuated the apparent degree of this overhang. On the right side, there is only a suggestion of the marginal horn-like rim seen in larger skulls. The mass of the boss, which resembles a flattened pillow, overhangs the subjacent prefrontal and frontal bones posteriorly (Fig. 13I). This is the position of the transverse tunnel seen in the Drumheller skull (Langston 1967). The superior surface of this boss is occupied by broad low hummocks, and a thick sagittal ridge rises above the general level of the top of the boss. The internal (ventral) surface of this boss is well preserved and is comprised of the paired nasal bones, whose sagittal suture remains open in this aspect. The ventral side of the boss is divisible into a low, broadly triangular central vault (Figs. 13C, 13D, 13E, 13G, 13H), a long excavation at the posterolateral edge of the vault, and a smaller, diagonally ovate excavation posterior to each of the posterolateral excavations. The surface of the large median vault is smooth and consists of cortical bone. It roofed the relatively wide posterior narial space. The adjacent excavations are separated from the central vault by sharp irregular sutural edges. They constitute the contact surfaces for attachment of the greatly thickened lacrimal in front and the prefrontal behind. That the contact surfaces for the prefrontals do not meet posteriorly at the midline suggests that an anterior wedge of the frontals made contact with the nasals beneath the posterior edge of the boss. The nature of the ventral sutural contact with the prefrontals precludes the entry of the frontals into this boss, at least. If this interpretation is correct, the elements in *P. canadensis* labeled as possible prefrontals by Langston (1975, fig. 4) may be frontals instead. A powerful ridge arises on either side of the midline within the nares. Extending forward, these ridges become the greatly thickened premaxillary processes of the nasals. The roofing bone of this boss is considerably thicker than in some larger skulls (Figs. 13G, 13H), and the sagittal suture is partly open. This boss is probably from a subadult individual because, uncoalesced, it has separated cleanly from the rest of the skull. It is, however, only a little smaller than some presumed adult examples. The differences may also reflect sexual dimorphism.

The supranasal boss (Fig. 6D) of *P. canadensis* has a massive, flattened, tumescent bulge on either side of a wide midlongitudinal sulcus immediately anterolateral to the nasal boss. This is absent in *P. lakustai*, although the opposing premaxillary processes of the nasal bones are thickened longitudinally to form a blunt-ended cylindrical mass. The external surfaces are smooth and are not textured like the surface of the nasal boss as they are in *P. canadensis*. The "fist-sized knob" (Fig. 6D) of *P. canadensis* (Sternberg 1950), a possible osteoderm, has not been identified in *P. lakustai*.

The posterodorsal rim of the external naris, which is formed by the nasal, flares laterally and is continuous ventrally with the distinctive long finger-like intranarial process (Langston 1975). This process is less strongly developed than in *P. canadensis* (Fig. 9C) but is more powerful than those of *Achelousaurus*, *Einiosaurus*, and other centrosaurines (Sampson 1995). The contact between the nasal and the premaxilla is a syndesmosis, in which a massive blunt-ended buttress on the medial side of the nasal rested upon a corresponding structure on the posterior ascending process of the premaxilla. The buttress was bounded medially by a thickened vertical flange that prevented the ascending process of the premaxilla from sliding or collapsing inward. In some skulls, this buttress and flange unit is 80 mm in transverse thickness. In contrast with some specimens of *Achelousaurus* (Sampson 1995), this union remains unfused even in the most massive skulls. Anteriorly, beneath the nasal boss, the nasal has a greatly thickened premaxillary process that forms the upper part of the narial bridge. Opposite processes are separated anteriorly by a triangular slot that received the united dorsomedial wedge-like processes of the premaxillae (Figs. 13B, 13F).

Eight detached nasal bones from immature individuals have been recovered. The smallest of these, a right nasal (TMP 1986.55.48, Fig. 15A), bears no resemblance to the massive structure of the adult nasal boss; had it been found in another context, it would surely have been attributed to some other centrosaurine that lacks a nasal boss. An almost complete left nasal (TMP 1988.55.80, Figs. 15B, 15C, 15D) is about two-thirds of adult size and is no more like an adult boss than the smaller example. This nasal has a low, anteroposteriorly broad-based, and laterally compressed demihorn (one-half of the sagittally divided nasal horn core) with a truncated top. The paired demihorns, fundamentally like those of "*Monoclonius*" and "*Brachyceratops*," appear narrow and fin-like from the front (Fig. 15D). The lateral surface of the horn is coarsely sculptured with sharply defined sinuous and anastomizing nutrient grooves and foramina. Below, the sculptured surface is gradational into the smoother lateral surface of the nasal. On the medial side, especially toward the dorsal edge, are seen the fine closely spaced striations characteristic of immature ceratopsid dermal bone.

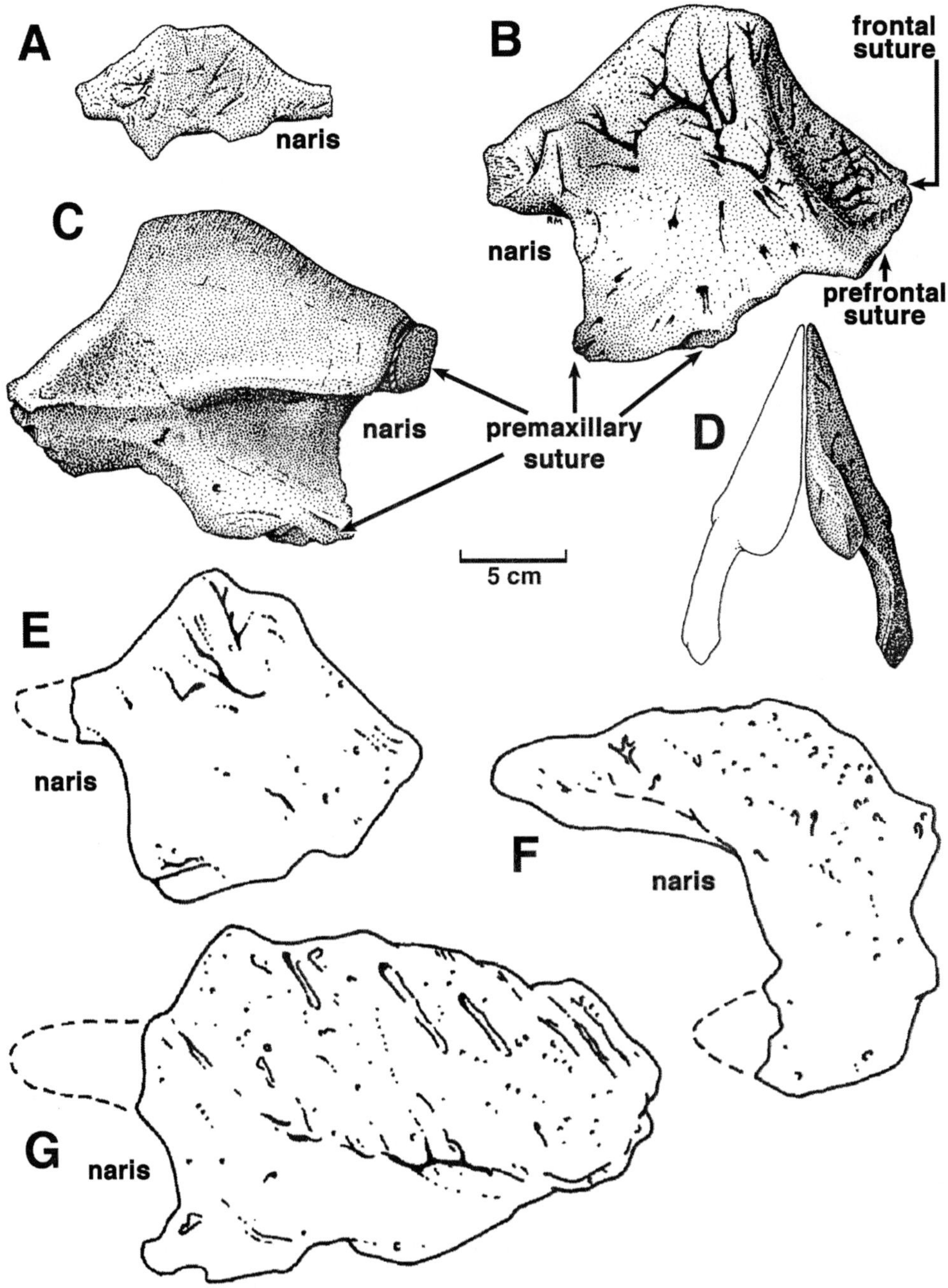

Fig. 15. Pachyrhinosaurus lakustai *nasal bones of juvenile to young adult individuals: (A) TMP 1986.55.48, the smallest example, right side (anterior to the right). (B, C, and D) TMP 1988.55.80 in left lateral (anterior to the left), left medial, and anterior aspects. (E) TMP 1987.55.161; (F) TMP 1989.55.256, reversed; and (G) TMP 1989.55.1342 in left lateral view.*

A pronounced change of plain occurs in the lateral surface of the nasal posteriorly along a sulcus from the edge of the nasal to the top of the horn. This results posteriorly in a flat, deeply sculptured triangular area that is obtusely deflected mesad. No such feature is seen in *Centrosaurus* or *Styracosaurus*, where the horns are transversely rounded. The sutures with the prefrontal and frontal are found at the base of the triangular area, suggesting that the frontal may have extended farther forward than in other centrosaurines (except

Achelousaurus) and (or) that the base of the horn extended farther posteriorly on the nasal than usual. Anteriorly, where the leading edge of the demihorn extends onto the narial bridge, the nasal bone expands slightly laterally and dorsally to produce a flattened bulge at the posterior end of the bridge. This feature is significant in the developmental history of the nasal boss, which will be discussed subsequently.

Medially, the immature nasals (Fig. 15C) are similar to those of *Centrosaurus*. There is no clear evidence of an internasal suture in any immature nasal bone, and most of the opposing bony surfaces of the demihorns apparently failed to contact each other. The posterior edge of the demihorn is beveled and irregular along its sagittal face, confirming that bony contact was not achieved posteriorly. There is no evidence of fusion occurring between demihorns in any of the immature *P. lakustai* specimens, although fusion in divided nasal horn cores commences at the tip and progresses downward in other ceratopsids. Below the demihorn, the lateral surface of the nasal bone is thin and platelike, displaying little flaring around the edges of the nares. Instead of the massive buttress and flange that contact the ascending process of the premaxilla in the adult nasal boss, there is only a tongue-and-groove contact like that in *Centrosaurus*. However, the beginnings of the broad syndesmotic surface of the adult nasal is discernible in TMP 1988.55.80 (Fig. 15C). The long finger-like intranarial process has only just begun to attenuate.

The small nasal (TMP 1986.55.48) has a relatively undeveloped demihorn (Fig. 15A), and the sculptured triangular posterior surface is less strongly deflected than in the larger specimen.

Somewhat larger right (TMP 1989.55.256, Fig. 15F) and left (TMP 1989.55.1111) nasals exhibit some thickening of the lateral walls. They also have some hypertrophied features, including the premaxillary process, the posterodorsal bulge of the narial bridge, and the intranarial process. The syndesmotic buttress for contact with the posterior ascending process of the premaxilla, although less developed than in skulls with nasal bosses, is much more massive than in the smaller nasals. What most strikingly distinguishes each of these larger nasals from the smaller ones is the presence of a broad tumid expansion of the area occupied by the demihorns in the smaller specimens. Distinctive features of the demihorns are no longer readily discernible, the structure now clearly foreshadowing the development of the nasal boss. The area corresponding to the base of the demihorn is considerably thickened transversely; a lateral ridge bounded medially by a wide depression probably corresponds to the superior edge of the demihorn.

An incomplete left nasal (TMP 1989.55.1342) is substantially larger than the others and has an obvious boss structure (Fig. 15G). Although as high as the boss in adult skulls, it is relatively short for its height and ends posteriorly at a more forward position on the skull roof than in more mature bosses. Fusion of the left and right nasals seems to have started anterodistally as in *Centrosaurus* (Ryan 2003) and other centrosaurines.

Supraorbital Unit

The orbit is surrounded by lacrimal, prefrontal, palpebral, postorbital, and jugal. All of these bones are separate elements in young individuals but co-ossify into a single supraorbital unit in the largest individuals. The frontal does not take part in the orbital margin in ceratopsids but forms part of the roof of the orbit and also fuses into the supraorbital unit. The sequence of fusion seems to have proceeded in the following order in most individuals: prefrontal and palpebral; postorbital and frontal; prefrontal–palpebral and postorbital–frontal; prefrontal–palpebral–postorbital–frontal and lacrimal; prefrontal–palpebral–postorbital–frontal–lacrimal and jugal. In old individuals, the contact between palpebral and postorbital is often still visible, even through the bones have fused together. The jugal has fused into this unit in only a few of the disarticulated specimens, including TMP 1989.55.367 (Fig. 16). The interfrontal suture between left and right supraorbital units never seems to fuse completely, which apparently is true for all centrosaurines (Ryan 2003).

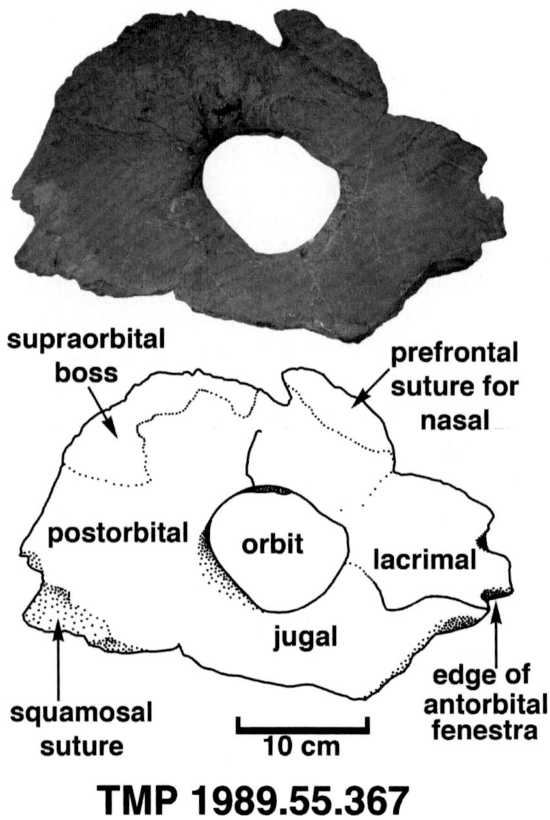

Fig. 16. Pachyrhinosaurus lakustai *(TMP 1989.55.367), right orbital region in lateral view.*

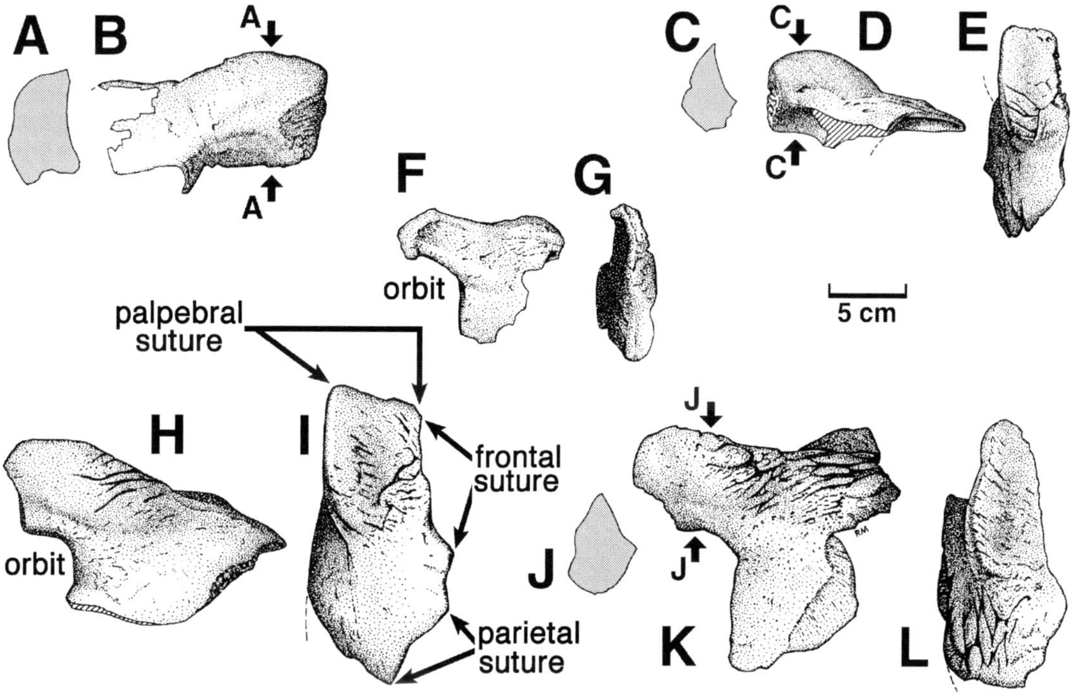

Fig. 17. Pachyrhinosaurus lakustai, *postorbitals. TMP 1986.55.155 in cross section (A) and right lateral view (B); TMP 1986.55.249 in cross section (C), left lateral (D), and dorsal (E) views; left lateral (F) and dorsal (G) aspects of TMP 1989.55.484; left lateral (H) and dorsal (I) views of TMP 1987.55.110; cross section (J), left lateral (K), and dorsal (L) views of TMP 1989.55.127.*

The supraorbital boss begins as a rather small, pointed horn on the postorbital in young individuals (Fig. 17). Like other centrosaurines, it rises vertically from the orbital rim, but the dorsomedial surface slopes more gradually towards the frontal suture. As the horn grows ontogenetically into a boss, it spreads by becoming thicker and more rugose medially until it extends onto the frontal, the back of the palpebral, and the posterolateral edge of the prefrontal as it does in *Achelousaurus* (Sampson 1995) and *P. canadensis* (Langston 1975). As in other centrosaurines, there is a remodeling stage in the largest individuals (Sampson et al. 1997; Ryan et al. 2001; Tanke and Farke 2006), and large supraorbital pits develop in positions formally occupied by the core of the supraorbital horns or bosses.

The orbital region of *P. lakustai* lacks the massive anterior buttress of the prefrontal and lacrimal that is so strongly developed in ceratopsids with supraorbital horns. The buttress is also lacking in *Achelousaurus*, *Einiosaurus* (Sampson 1995), and *P. canadensis,* but is present in *Centrosaurus*.

Lacrimal

The lacrimal has been only partly defined in previously described *P. canadensis* skulls; sutures drawn by von Huene (1950) on a sketch of a *P. canadensis* skull (CMN 8867) are largely fanciful. Dissociated left lacrimals (Fig. 18) from subadult individuals (TMP 1988.55.181, TMP 1989.55.10, TMP 1989.55.20) resemble that of *Centrosaurus* and display similar relationships to adjoining elements. It is more massive than in other ceratopsians, including *Centrosaurus*, although the bone comprises less than one-quarter of the rim of the orbit. As in *Achelousaurus*, *Einiosaurus*, and *P. canadensis* but unlike chasmosaurines, it is thickened chiefly by the addition of bone on its medial side and is essentially featureless laterally. The bone is traversed by several deep, narrow subparallel radiating sulci that pass anteroventrally from inside the orbit to emerge onto the lateral surface (Fig. 18A), where they impart a wrinkled appearance to the edge of the orbit. One prominent groove continues anteriorly for most of the length of each of the lacrimals (Fig. 18A). No crater-like excavation as occurs in *Achelousaurus* is present on the side of the lacrimal. The lacrimal duct passed through a notch (TMP 1989.55.20) on the medial edge of the orbital rim, and entered the upper of two shallow longitudinal excavations that divide the medial surface of the bone.

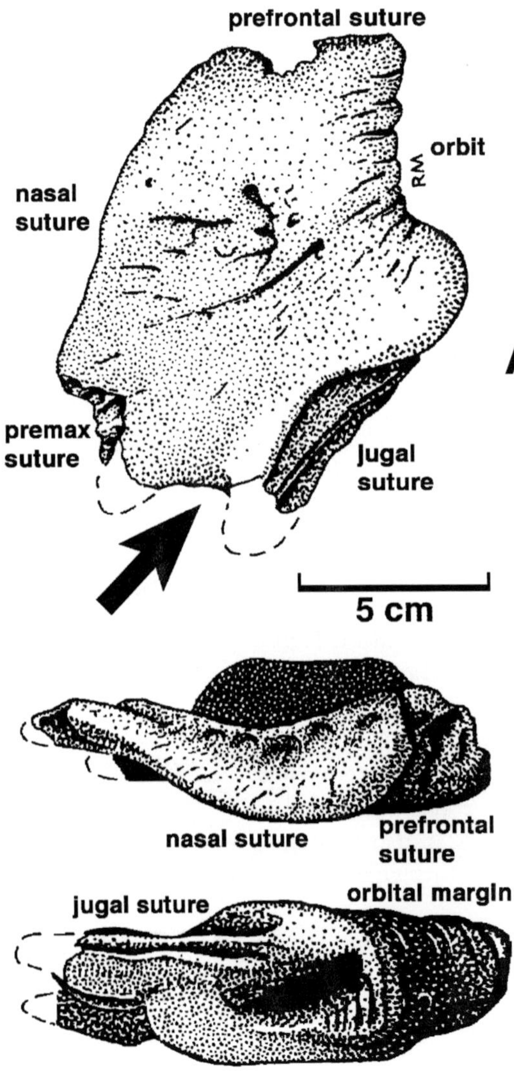

Fig. 18. Pachyrhinosaurus lakustai, *left lacrimal (TMP 1988.55.181) in lateral (A), anterodorsal (B), and posteroventral (C) views. Arrow indicates lacrimal segment of the antorbital fenestra.*

A short suture for the prefrontal is formed of deeply interdigitating grooves and ridges of a sutura limbosa. The contact with the premaxilla is a combination of scarf and butt joints that allowed the narrow ascending ramus of the premaxilla to grow and prevented backward dislocation of this bone under the stresses generated by movements of the jaw. The long contact with the nasal (Fig. 18B) expands transversely posteriorly and comprises both lap and butt joints. In the broadest area of contact between the lacrimal and the nasal, the suture is strengthened by transverse ridges in larger specimens (TMP 1989.55.367,

Fig. 16). Anteriorly, the ventral edge has two long, narrow longitudinal grooves (Fig. 18C). The more lateral of these grooves is shallow, and its medial side, which is broadly exposed in lateral aspect (Fig. 18A), formed a lap joint and shallow schindylesis (wedge-and-groove joint) with the jugal. Medial to this contact, the lacrimal has another longitudinal groove for the jugal. This is continuous posteriorly with a broad medially expanded flange that has a combination of butt joint and interlocking grooves and ridges. The central part of this area includes a large gomphosis (peg-and-socket joint) with the jugal (Fig. 18C). These complicated joints permitted ontogenetic enlargement of the orbit, helped distribute stresses emanating from the nasofrontal region, and restricted deformation of the side of the face and orbit.

A broadly rounded notch in the anteroventral edge of the lacrimal forms the posterodorsal margin of a small antorbital fenestra, which has also been called the preorbital fossa or the intraorbital foramen. Occurrence of this opening is variable in *P. canadensis* as it is in other large ceratopsids (Dodson and Currie 1990). It is small in *P. canadensis* (and is apparently absent in the Drumheller skull) and may have been closed in some of the Pipestone Creek specimens. However, the posterodorsal margin is preserved on all isolated lacrimals and in TMP 1989.55.367 (Fig. 16B). It is a large, narrowly oval fenestra on either side of the face in TMP 1987.55.285, measuring 70 mm long and 16 mm wide on the left side. The fenestra is small in neoceratopsia, and is subject to considerable variation in pachyrhinosaurs.

Prefrontal

The prefrontal bone is generally unrecognizable in adult *P. canadensis* skulls, but detached prefrontal bones have been recovered in *P. lakustai*. Langston (1975) suggested that the united prefrontals comprise more than one-half of the mass of the nasal boss. The prefrontal does not extend forward to the nasal boss in *Achelousaurus* where that boss (Sampson 1995, fig. 5) is relatively smaller than in adult *P. canadensis*.

There are several isolated prefrontals (TMP 1989.55.347 (right), TMP 1989.55.1198 (left), TMP 1989.55.178 (right, fragment)), and two (TMP 1989.55.1596 (left) and TMP 1989.55.1198 (left)) partially fused to palpebrals (Fig. 19). In TMP 1989.57.367 (Figs. 16, 20A) and larger specimens, the prefrontal is fused to frontal, postorbital, lacrimal, and palpebral, although some of the sutures can still be discerned. Viewed from above (Figs. 19A, 19B), the prefrontal appears as a transversely elongate trapezoid whose medial boundary with the frontal bone is more than twice as long as the lateral side where it forms part of the orbital margin and has a suture with the lacrimal. The prefrontal is narrowly exposed on the orbital rim for a distance of 11 mm in TMP 1989.55.178. In TMP 1989.55.1596 (Figs. 19A,

Fig. 19. Pachyrhinosaurus lakustai. *TMP 1989.55.1596, left palpebral and prefrontal, in dorsal (A) and ventral (E and F) views; TMP 1989.55.347, left prefrontal in dorsal (B) and ventral (C and D) views.*

19E, 19F), it has a 10 mm exposure on the orbital rim at its narrowest point, whereas the lacrimal suture of TMP 1989.55.347 (Figs. 19B, 19C, 19D) is only separated by 6 mm from the palpebral suture. The dorsal surface is shallowly excavated anteroposteriorly in these smaller individuals, becoming more concave laterally toward the junction with the lacrimal. The surface is traversed by numerous broad ridges and narrow intervening grooves trending more or less transversely. A

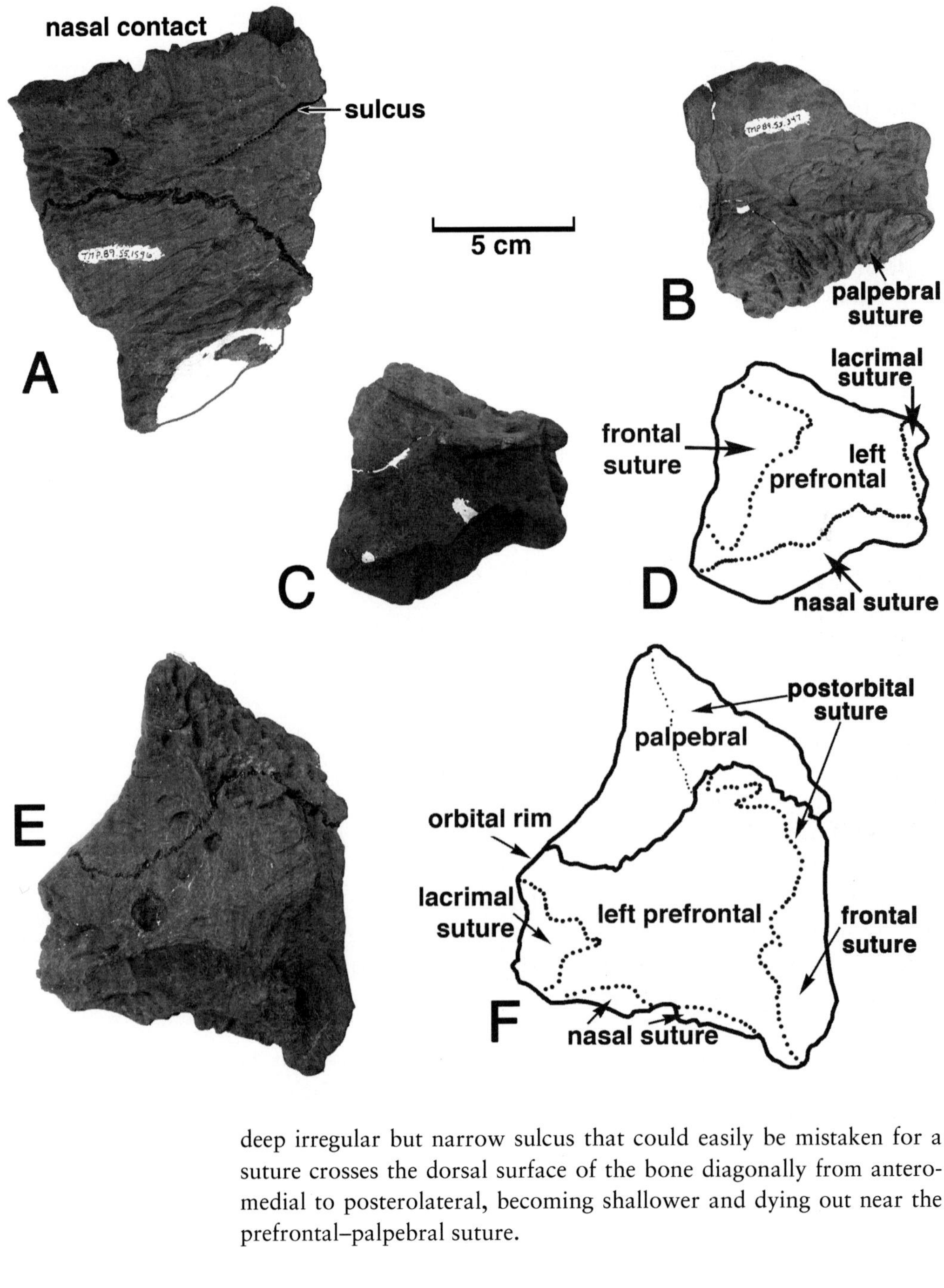

deep irregular but narrow sulcus that could easily be mistaken for a suture crosses the dorsal surface of the bone diagonally from anteromedial to posterolateral, becoming shallower and dying out near the prefrontal–palpebral suture.

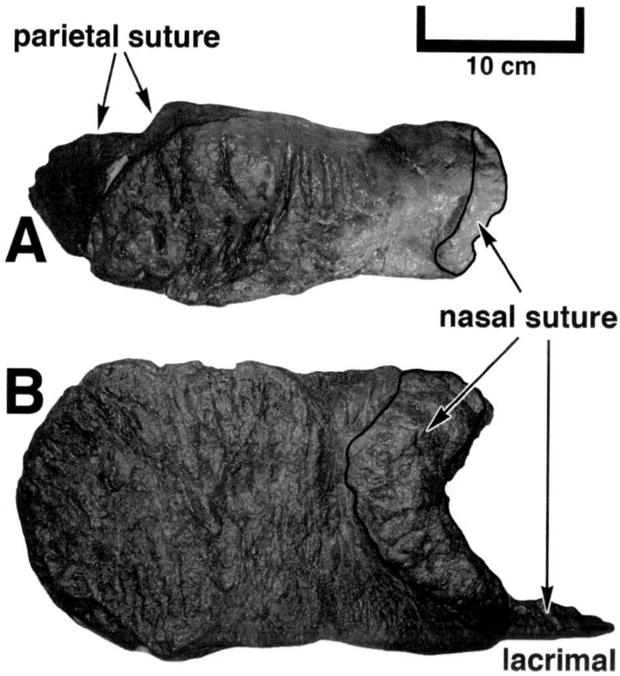

Fig. 20. Pachyrhinosaurus lakustai *supraorbital bosses (fused postorbitals, palpebrals, prefrontals, and frontals) in dorsal view: (A) TMP 1989.55.1131; (B) TMP 1989.55.367.*

The sutural surfaces of the prefrontal are complex. That between prefrontal and lacrimal consists of four to seven deep, roughly vertical grooves that received corresponding ridges from the lacrimal (sutura limbosa). The overlapping contact with the frontal is shallow in TMP 1989.55.347 (Figs. 19E, 19F) and is less than one-half the height of the lacrimal suture. However, it has become relatively deeper in the larger TMP 1989.55.1596 (Figs. 19C, 19D) and is as almost as thick as the lacrimal suture. It is clear from the larger co-ossified orbital masses that the thickness of the prefrontal continued to increase with strong positive allometry. The sutural surfaces for the frontals are little roughened and display qualities of the squamous sutures in the smaller specimens. This ventral surface slants dorsomedially at an angle of about 45° as it overlaps the edge of the frontal. Lateral to the frontal suture, there is a strong transverse ridge (Figs. 19C, 19E) on the ventral surface that extends laterally to the lacrimal suture and medially onto the frontal. Three to five irregular but well-defined deep pits are always found on the ventral surface of the prefrontal posterior to this ridge (Figs. 19C, 19E). The contact with the nasal (Fig. 20) consists of a relatively wide, deeply marked surface with multiple socket-like excavations on the ventral surface that evidently received corresponding projections of bone from the nasal. These excavations are inclined upward and forward at an angle of about 45° to the surface. Lateral to this powerful region of contact is a more limited, virtually smooth surface. In the smaller specimen (TMP 1989.55.347, Figs. 19C, 19D), the surface is oriented ventrally, and the bone is dorsoventrally thin. The slightly larger specimen (TMP 1989.55.1596, Figs. 19E, 19F) is significantly thicker in this region, which has become a dorsoventrally thick surface that marks the area where the massive nasal boss has begun to override the prefrontal (Fig. 19A). In TMP 1989.55.98, the suture with the nasal is concave in dorsal view, and forms a shelf that overhangs the back of the nasal. The nasal fits into a horizontal slot, below which the prefrontal again forms a shelf on which the nasal sits. Medially, this shelf develops into a prominent triangular process that plugs into a conspicuous triangular slot in the ventral surface of the nasal boss (TMP 1986.55.206). The prefrontal becomes massively thick in TMP 1989.55.367 (Figs. 16, 20B), TMP 1989.55.808, TMP 1989.55.1131 (Fig. 20A) and other specimens, where it forms a ridge that is up to 10 cm thick medially. In dorsal view (Fig. 20B), this ridge is concave anteriorly and has reinforcing anteroposterior ridges and grooves that strengthened the contact with the nasal boss. At this stage, the nasal boss presumably would have obscured the prefrontal suture in dorsal aspect. Medially, the ridge is continuous with an upturned ridge at the anterior end of the frontal that also supports the underside of the nasal boss. In its ultimate expression in *P. canadensis*, this overriding process of the nasal obscures the prefrontal in dorsal view. However, the bone persists to form the floor of the transverse tunnel that crosses the skull roof between the nasal boss and the supraorbital bosses in the Drumheller skull (Langston 1967), a condition that is also inferred in a skull (CMN 9485) from Scabby Butte (Langston 1975). However, the tunnel is not present in any of the *P. lakustai* skulls, because the nasal boss never extends far enough back to contact the supraorbital bosses.

The combined widths of the prefrontal bones in TMP 1989.55.1596 (Fig. 19A) would have been 223 mm, which is less than the width of the skull roofs in any subadult skull of *P. lakustai*, suggesting that the prefrontals did not meet on the midline at this stage of growth. The squamous sutural surfaces on the prefrontal (Figs. 19D, 19F) suggest that this bone would have overridden the frontal in life, although this is not clearly seen in any of the specimens because of fusion. It seems possible that *P. canadensis* also retained the primitive state of no prefrontal to prefrontal contact (as do other centrosaurines except *Styracosaurus* and *Centrosaurus*), although this cannot be determined with certainty (Sampson 1995).

Palpebral

The existence of a palpebral element (supraorbital of some authors) in ceratopsians has long been equivocal (Coombs 1972), even though a freely articulating palpebral is present in Psittacosauridae (Sereno 1990) and Protoceratopsidae (Brown and Schlaikjer 1940). Although it is usually obscured in massive ceratopsid skulls, a palpebral that is integrated into the roof of the orbit is sometimes visible in immature individuals, and its universal presence in the Ceratopsia now seems certain.

The palpebral bone of *P. lakustai* is represented by an isolated, well preserved subadult example, TMP 1988.55.105 (Fig. 21). Another specimen (TMP 1989.55.1596) of comparable size is joined to the prefrontal, but the suture between the bones is visible (Figs. 19A, 19E, 19F). The palpebral forms the rim of the orbit between the prefrontal and postorbital and excludes the frontal from the rim. It thickens posteriorly (Fig. 19C) to reinforce the anterior part of the base of the postorbital horn. Viewed from above, the bone is a massive, almost equilateral, wedge-shaped element. Except for the orbital rim, its margins are occupied by coarse sutural grooves and ridges (sutura limbosa). The superior surface is marked by a few wide and many narrow, roughly transverse and sometimes anastomosing sulci and by a few foramina. Laterally, at about midlength, the dorsal edge is emarginated by a broad-bottomed transverse excavation that becomes shallower and disappears medially about one-third of the way across the superior surface of the bone. This trough (Figs. 21A, 21C) may persist in adults as a strong transverse groove between the nasofrontal and supraorbital bosses in *P. lakustai*, *P. canadensis*, and *Achelousaurus*. The area of the palpebral exposed within the roof of the orbit is only about one-fourth of the area of the superior surface, and the bone functioned like a keystone wedged between the prefrontal, postorbital, and frontal bones. It is separated from the lacrimal in TMP 1989.55.1596 by a thin process of the prefrontal that reaches the orbital rim.

Seen from above, the palpebral is roughly triangular as in *Achelousaurus* (Sampson 1995), *Einiosaurus* (Sampson 1995), "*Monoclonius lowei*" (Sternberg 1940) and mature specimens of

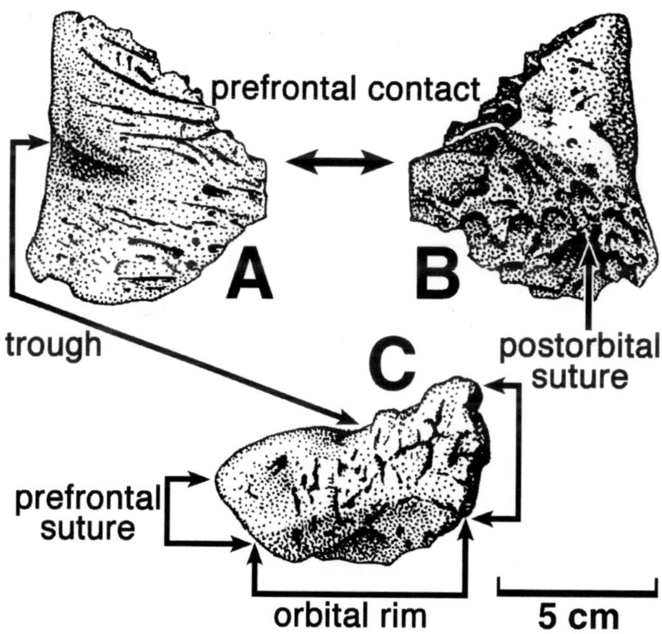

Fig. 21. Pachyrhinosaurus lakustai, *left palpebral (TMP 1988.55.105) in dorsal (A), ventral (B), and lateral (C) views. The double-headed arrow shows the junction where the palpebral, prefrontal, and postorbital meet.*

Styracosaurus (TMP 1999.55.1; Sternberg 1927). The palpebrals of a subadult specimen of cf. *Styracosaurus* (TMP 1998.93.64) and most specimens of *Centrosaurus* (TMP 1980.54.1) are more rounded and elongate. Whereas the dorsal surface is relatively flat in *P. lakustai*, the palpebral of *Centrosaurus* and the juvenile cf. *Styracosaurus* curves anteroventrally to help form a conspicuous raised ridge (antorbital buttress) on the anterodorsal orbital margin (Sternberg 1940; Sampson 1995). As in other large ceratopsids, the bone is definable in adult skulls only by its topographic position, owing to its fusion with adjacent elements. The palpebral became thicker medially and posteriorly but, apparently, was not incorporated into the supraorbital boss (TMP 1989.55.808 and TMP 1989.55.1131). The dorsal anterolateral edge of the palpebral TMP 1989.55.1596 is rolled upward slightly, with a row of four small hummocks dorsolaterally, suggesting the rudiments of a supraorbital horn, but the small hornlike structures seen in *Achelousaurus* (Sampson 1995) and *P. canadensis* (including the Drumheller skull) are not developed in any *P. lakustai* skull.

Frontal

Frontal bones are difficult to define in large ceratopsid skulls, where they fuse into the postorbital and prefrontal. The distinctive frontal fontanelle, which is relatively small in *Centrosaurus* and *Styracosaurus* (Sampson 1995), extends into a sinus system (supracranial cavity) that invades the supraorbital horn or boss in *P. lakustai*, *Einiosaurus*, and *P. canadensis*. This indicates that there is considerable resorption and restructuring inside the frontal during ontogeny. Isolated frontals, such as TMP 1989.55.369 (Fig. 22), are rare.

Fig. 22. Pachyrhinosaurus lakustai, *TMP 1989.55.369, right frontal in medial (A), dorsal (B), and lateral (C) views. The medial view shows a series of sinuses, whereas a strong ridge-and-groove prefrontal suture is visible in lateral view.*

The frontals took no part in the nasal boss in *P. lakustai* and *P. canadensis*. The frontals are overridden anteriorly by posterior expansion of the nasals in *P. canadensis* (Langston 1975) and *P. lakustai* (TMP 1989.55.367, Fig. 16; TMP 1989.55.808, Fig. 23C; TMP 1989.55.1131, Figs. 23A, 23B). Dimensions and sutural edges on the prefrontals of *P. lakustai* show that these bones were separated medially by the frontals. In a large *P. lakustai* skull (TMP 1986.55.111, Fig. 8E) in which the "sponge-shaped" supraorbital bosses are well developed, the frontal forms part of the medial side of each boss, similar to *Achelousaurus* and *P. canadensis*. Dorsally, the frontal extends posteriorly to overlap the anterolaterally directed ramus of the parietal, thereby excluding the postorbitals from the midline (Sampson 1995). Nevertheless, in stages prior to the appearance of the supraorbital boss, the frontal is joined posterolaterally to the edge of the shield-like postorbital bone (TMP 1987.55.47).

The unique and enigmatic ceratopsid frontal fontanelle is keyhole shaped (TMP 1986.55.111, TMP 1989.55.808). One of the best preserved examples is in TMP 1986.55.111, a massive skull (Fig. 8E) that closely resembles *P. canadensis* from Scabby Butte. Here, the fontanelle is 190 mm long, wider (30 mm) behind than in front (15 mm) and narrowest (7 mm) between the large supraorbital bosses. Posteriorly, and for a short distance on either side, the fontanelle is bounded by the anteromedial Y-shaped extension of the parietal bar. The frontal extends forward to form a narrow overhanging edge of the fontanelle medial to the supraorbital boss. In this skull, the fontanelle is bounded by the frontal and parietal, as is usual in ceratopsids (Sampson 1995).

The fontanelle ends anteriorly between the supraorbital bosses. Farther forward, between the supraorbital bosses and 25 mm in front of the fontanelle, the skull roof of TMP 1986.55.111 has a broad funnel-shaped excavation that extends downward 35 mm below the supraorbital cranial roof. The mouth of this opening is 90 mm wide and 40 mm long. Its lateral edges lie about 80 mm from the dorsolateral margin of the supraorbital skull roof. Whether this funnel-like depression joins the frontal fontanelle beneath the narrow bridge of bone between the anterior ends of the supraorbital bosses is not known. During ontogeny, the frontal fontanelle normally closes by mesial expansion of its lateral (frontal) edges, beginning anteriorly and progressing caudad (Sampson 1995). The large *P. canadensis* skull from Scabby Butte (CMN 9485) displays intermittent bridges across the fontanelle, however, so the 25 mm wide expanse of bone separating the funnel-shaped depression and the fontanelle in TMP 1986.55.111 may represent one such bridge, the

Fig. 23. Pachyrhinosaurus lakustai, *frontals in medial view of supraorbital bone complex. (A and B) TMP 1989.55.1131; (C) TMP 1989.55.808.*

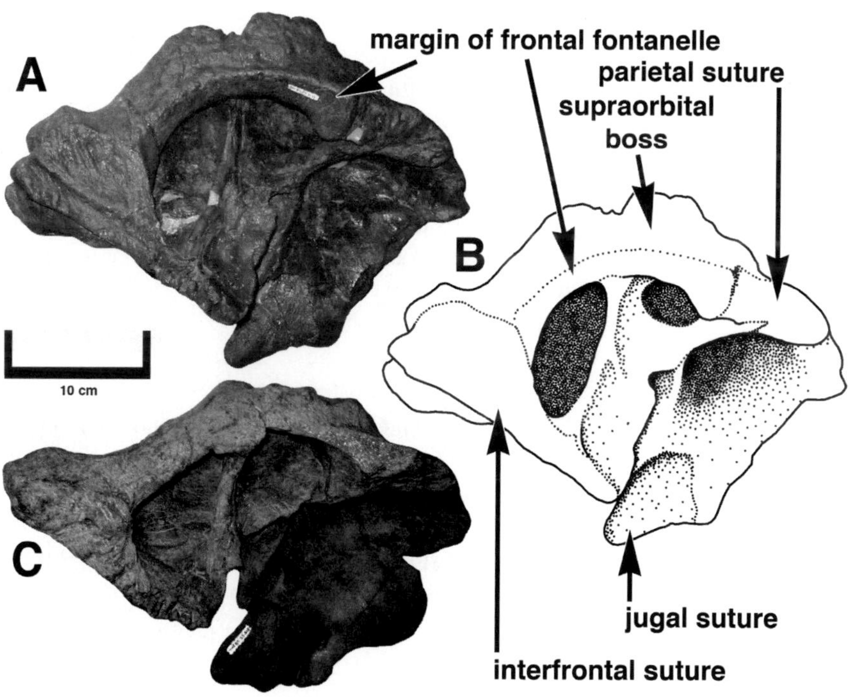

depression being merely the anterior end of the fontanelle. This condition has not been observed in any other skull from Pipestone Creek, and it is not present in the Drumheller skull.

The frontals do not fuse to each other until after the circumorbital ring of bones co-ossifies into a single unit. However, the contacts between the frontals deepen ontogenetically especially anterior to the frontal fontanelle. Whereas the frontal initially appears to have been not much thicker dorsoventrally than the prefrontal, a posteroventral process of the interfrontal suture increases dramatically in length as the lateral wall of the frontal fontanelle deepens. This triangular, ventrally tapering process extends almost 10 cm below both the roof of the frontal fontanelle and roof of the orbit in TMP 1989.55.808 (Fig. 23C) and 15 cm below in TMP 1989.55.1131 (Figs. 23A, 23B). The interfrontal suture extends down the medial surface of the anteroventral edge of this process so that the frontals floor the anteroventral part of the frontal fontanelle. The interfrontal suture is deeply ridged in these two specimens, and it appears likely that the interfrontal suture was fused in larger specimens. As in *Centrosaurus* (TMP 1998.93.64), the lateral region of the frontal fontanelle is divided into two deep pockets by an anteroventrally sloping ridge. These pockets may be further subdivided by smaller ridges in some specimens.

The frontal fontanelle is continuous with large supracranial cavities that extend laterally and undermine the supraorbital boss. These cavities are found in other ceratopsids; however, in *Centrosaurus* and *Styracosaurus*, they do not invade the bases of the supraorbital horn cores.

Postorbital

The postorbital bone was only loosely integrated into the skull roof until the supraorbital boss reached an advanced stage of development; dissociated postorbitals are the most frequently found skull bones in the Pipestone Creek collection, and they provide an extensive ontogenetic series (Figs. 17, 24). Like the nasals, the postorbital bone in its adult state is so different from the juvenile condition that, if found separately, examples of the two growth stages would surely be taken for different taxa. Prior to the subadult stage, postorbitals resemble the bone in the juvenile skulls of "*Brachyceratops montanensis*" or subadult *Einiosaurus* (Sampson 1995). As in those specimens, a low, supraorbital horn rises above the rim of the orbit posterodorsally (Fig. 17). The original horn core is retained in recognizable form in subadults, but transverse ridges develop medial and posteromedial to the center of the horn as these parts of the postorbital and probably the frontal thicken into a series of transverse ridges also seen in *P. canadensis* (Langston 1967, 1975), *Achelousaurus*, and *Einiosaurus* (Sampson 1995). The horn core is long based and ridge-like, ranging 50–60 mm in height above the orbit in available specimens. These

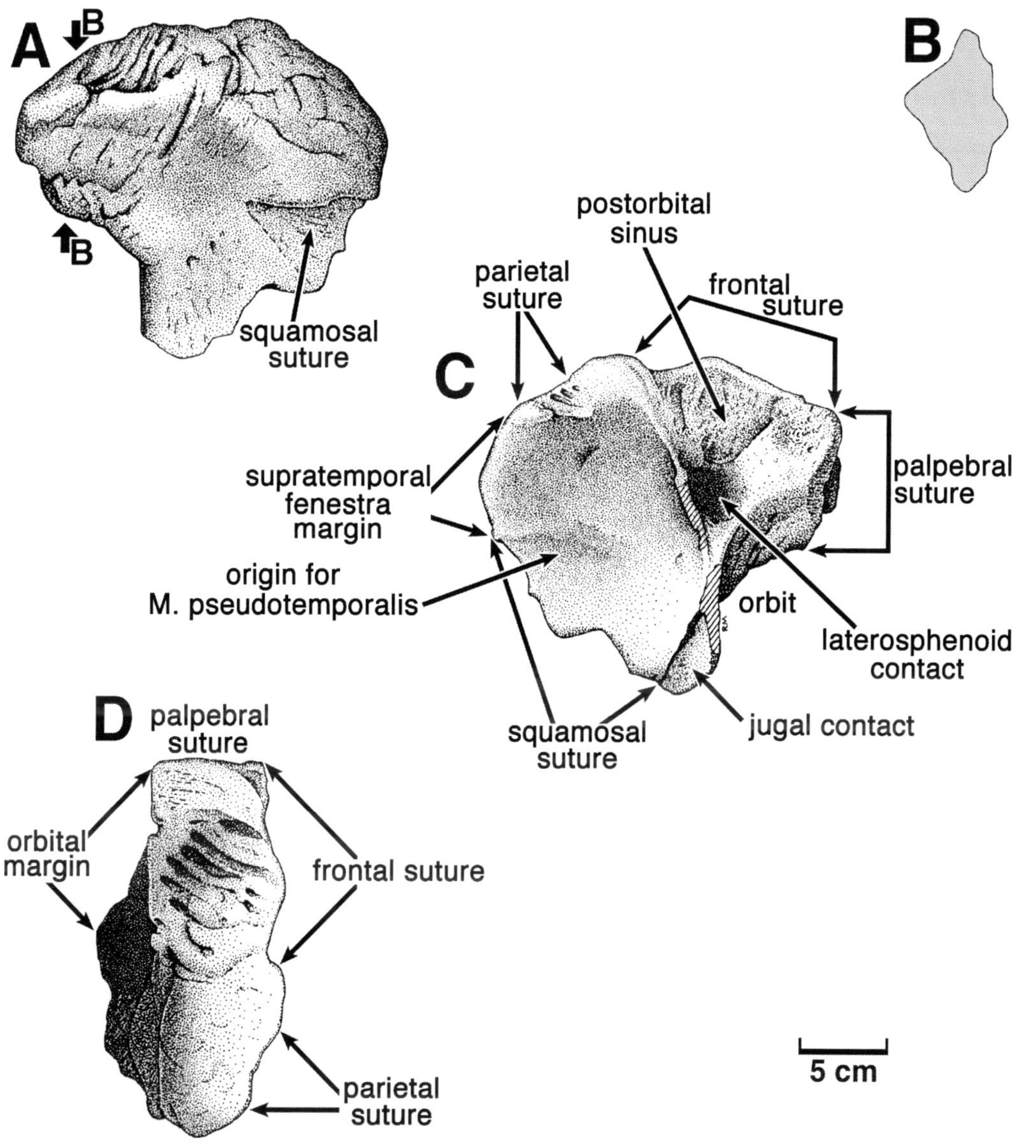

Fig. 24. Pachyrhinosaurus lakustai, *left postorbital of subadult individual, TMP 1987.55.47, in lateral aspect (A), cross section (B), and medial (C) and dorsal (D) views. In Fig. 24D, the anterior edge is up, and the lateral surface is to the left.*

horns are roughly triangular in transverse section; as in subadults of *Achelousaurus* (but unlike *Einiosaurus*), they are slightly excavated dorsoventrally on the medial side (Fig. 24D). Viewed from the front, their longest transverse axis is inclined upward and outward, so the lateral side extended, brow-like, over the orbit to a limited extent. Behind the postorbital horns (or, in adult skulls, the boss), the postorbital bone is broadly convex and shield-like. There is no superficial crater-like excavation as is seen laterally on the postorbital of *Achelousaurus horneri*. Posteriorly and ventrolaterally, the postorbital makes overlapping and underlapping squamous contact, respectively, with the jugal and the squamosal. The squamosal suture is bounded proximally by a schindylesis of considerable depth.

Medially, the postorbital displays a sutural surface showing that, as in other centrosaurines, this bone was excluded, at least anteriorly, from the edge of the frontal fontanelle by the frontals (as first noted in *Styracosaurus* by Sternberg (1927); see also Lehman (1990)). However, in TMP 1987.55.47 (Figs. 24C, 24D), what appears to be a short finished edge occurs between the suture for the frontal and that for the parietal, indicating that the postorbital entered the edge of the fontanelle briefly in this individual. Whether this is a frequent condition in *P. lakustai* is unknown. (Sampson (1995) believes that the postorbital never enters the fontanelle in centrosaurines—but see Dodson and Currie (1990).) The anterior edge of the postorbital is a broad deeply corrugated sutural face (a sutura harmonia of classical anatomy), which joined the palpebral bone, thus excluding the frontal from the orbital rim.

Specimen TMP 1987.55.47 (Fig. 24) is the largest of the detached postorbitals. It is almost as large as the postorbital in some of the adult skulls and is presumed to be from a subadult individual. The supraorbital boss is in an early stage of differentiation, but a remnant of the juvenile horn is still evident in a thickened hummocky dorsolateral ridge that rises some 75 mm above the orbit. Anteriorly, it appears to be in the process of "overriding" the palpebral bone. Posteromedially, the concave surface of the preexisting juvenile horn has been engulfed by the thickening of the bone above the orbit so that the top of the skull is now almost flat to slightly inclined upward medially. Bordering this inclined tabular surface posteriorly is a subround area approximately twice as large as the tabular area. Its upper surface (Fig. 24D) displays the diagonal ridges and grooves described previously in the section on the supraorbital boss. This "ridge-and-groove field" ends posteriorly where the ridges appear to pass beneath a shelf about 4 mm thick, which is contiguous with the hummocky sculptured postorbital surface behind and which may have originally extended over the ridge-and-groove field.

As the postorbital grew into the supraorbital boss, it incorporated the prefrontal and palpebral anterolaterally, and the frontal was incorporated into its medial edge as in *P. canadensis* (Langston 1975).

The inner side of the postorbital (Fig. 24C) has a rounded socket behind the posterolateral edge of the orbit that lodged the cartilage-capped tip of the dorso-anterolateral buttress of the laterosphenoid. Bounding this socket and passing ventrally is a massive ridge that helped to brace the skull roof against the braincase. This ridge forms the anterior rim of a broad, well-defined triangular space possibly occupied in life by the origin of the M. pseudotemporalis. Another triangular space dorsomedial to the socket for the dorso-anterolateral buttress of the laterospenoid represents the roof of the frontal sinus.

The "transverse tunnel" produced by overgrowth of the postorbital exostosis described in the Drumheller skull (Langston 1967) is undeveloped in *P. lakustai*. Whether this difference is systematically important or is a result of differing ontogenetic stages is unknown.

Supraorbital boss

The flattened, circular supraorbital boss (Figs. 4, 6, 7, 8, 13, 16, 20, 23, 24) was formed mostly of the postorbital bone; however, like the boss of *Achelousaurus* (Sampson 1995), it contains prefrontal, palpebral, and frontal components. In its mature state, the boss of *P. lakustai* resembles that of *P. canadensis*, but it occupies a relatively smaller area of the skull roof in *P. lakustai* and does not contact the nasal boss anteriorly. In skulls of *P. canadensis* from Scabby Butte and Drumheller, the top of the boss is flat to slightly convex, and the bone is several centimetres thick at its center. In all Pipestone Creek specimens so far prepared, the top of the boss is depressed, and the roofing bone is thin comparable with conditions in the holotype of *P. canadensis* (Sternberg 1950, pl. 17) but differing from Scabby Butte examples. Some exhibit strong transverse ridges and deep intervening grooves dorsally (the ridge-and-groove field) that traverse the boss diagonally from anterolateral to posteromedial. In *P. canadensis* skulls from Scabby Butte, similar features seem to be associated with the palisaded topography around the periphery of the bosses. The tops of the supraorbital bosses in TMP 1986.55.111 (Fig. 8) and other specimens are dorsally concave, whereas the floor of this concavity in TMP 1989.55.427 is pierced to enter the sinus underneath. These concavities are characteristic of large individuals of most centrosaurine species (Tanke and Farke 2006), including *Centrosaurus* and *Styracosaurus* (CMN 344), and represent resorption pits that develop at advanced age. The ridge-and-groove field probably represents a region of rapid growth, mobilized at a late juvenile stage, and development of the supraorbital bosses likely followed the same sequence as the nasal boss. The ridges developed first with the spaces between them being filled in later, producing a multilayered plicated pattern whose axis of breadth lay normal to the upwardly facing top of the boss.

Although the supraorbital boss looks massive, the ventral surface is excavated in mature animals by huge frontal sinuses that are continuous with the frontal fontanelle.

Temporal Region

The temporal region is composed of four bones—jugal, epijugal, quadratojugal, and quadrate. The squamosal is also involved in this region but is more appropriately described as part of the frill.

Jugal

There are several complete jugals and a number of incomplete examples of disparate size in the Pipestone Creek collection. A complete jugal is preserved with partial skull TMP 1989.55.188 (Fig. 8B). The best, disarticulated specimen (Figs. 25A, 25B) is the right jugal of a subadult individual (TMP 1987.55.88), which is about the size of the jugal of an adult *Centrosaurus*.

The thin, flat jugal seems a little broader than in *Centrosaurus* (Fig. 25C). The ventral edge is more obtuse, and the border of the lateral temporal fenestra, which is situated higher up on the jugal, is relatively shorter, reflecting the small size of this opening as in *P. canadensis* (Langston 1967, 1975). However, the orbital rim is longer than in *Centrosaurus* relative to the area of the bone. A vertical ridge that arises laterally and continues onto the epijugal in a number

Fig. 25. Jugals of Pachyrhinosaurus lakustai *and* Centrosaurus apertus. *(A and B) Right jugal of a subadult* Pachyrhinosaurus lakustai *(TMP 1987.55.88) in lateral and medial views; (C) medial view of right jugal of* Centrosaurus *(TMP 1980.18.115); incomplete left jugal of immature* Pachyrhinosaurus lakustai *(TMP 1989.55.149) showing partly ankylosed epijugal in lateral (D) and anterior (E) views.*

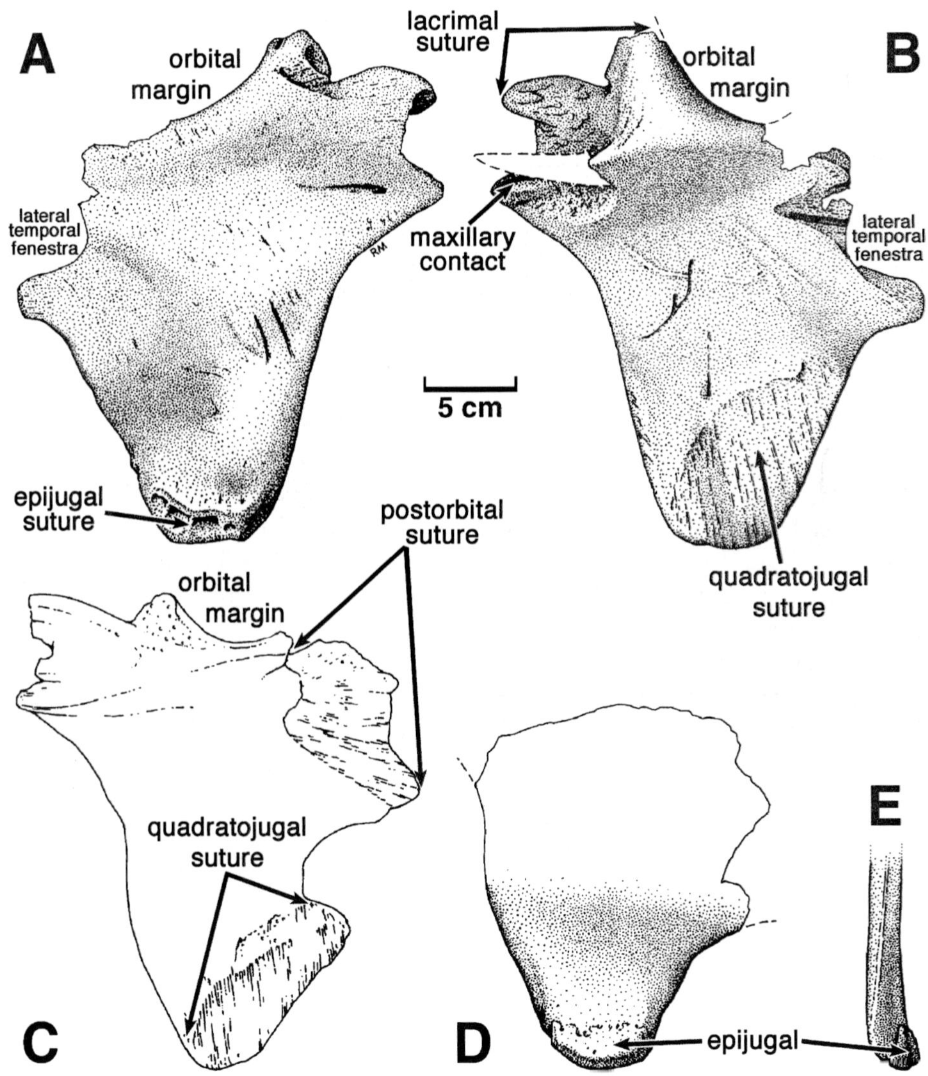

of other ceratopsians is absent; on the contrary, this area of the bone is slightly concave in *P. lakustai*. The thickening of the medial side of the orbital rim is considerably greater than in *Centrosaurus* (including "*Monoclonius*"), no doubt related to the hypertrophy of the cranial bosses in pachyrhinosaurs. As in the lacrimal, the rim is not expanded outward. The part of the jugal that joined the lacrimal is a complex set of grooves, ridges, and flat surfaces reminiscent of the interlocking but flexible arrangement in hadrosaurs (Langston 1960). However, the structure was nonfunctional in adult ceratopsians where jugal and lacrimal are firmly coalesced.

Except for sutural scars for the lacrimal, maxilla, quadratojugal, and squamosal and a set of broad reinforcing ridges near the orbit, the medial side of the jugal is featureless. There is no indication of muscle attachment except, possibly, a limited strip along the medial side of the anterior edge for the postulated origin of a cheek muscle, M. levator anguli oris. Tyson (1977, fig. 12) provides a photograph of *Centrosaurus* (UALVP 16248) that displays coarse striations on the medial side of the jugal that are interpreted by her as the origin for this muscle. However, similar striations in *P. lakustai* are clearly part of the attachment area for the quadratojugal, and we believe this is the explanation of the striations in Tyson's *Centrosaurus*. The striations are coarser than would be expected for a muscle attachment and, lying on the internal surface of the bone, are less well situated for the origin of a cheek muscle than is the anterior edge of the jugal. The sutural surface for the quadratojugal is substantially broader than in *Centrosaurus*.

The external surface of the jugal is sculptured by a profusion of shallow, closely spaced grooves seemingly without pattern. The ventrolateral edge of the jugal TMP 1987.55.88 is slightly thickened, roughened, and deflected outward. Four small rectangular pits near the edge mark the area where the epijugal was attached (Fig. 25A).

Epijugal

Fused epijugals are evident in several specimens. A small epijugal bone is incompletely ankylosed to the ventrolateral end of an incomplete jugal, TMP 1989.55.149 (Figs. 25D, 25E). It is small and lozenge shaped; unlike epijugals in most ceratopsians, its longest diameter is more anteroposterior than dorsoventral. It is similar to the epijugal recovered with a fragment of the jugal (TMP 1993.29.4) of the Drumheller skull. In one of the largest epijugals (TMP 1986.55.304), the tip of the bone has been resorbed in a manner similar to the nasal and supraorbital bosses (Tanke and Farke 2006). TMP 1985.112.90 is a relatively large (about 8.5 cm wide at the base), triangular epijugal that appears to have been only partially fused to the jugal at the time of death.

Quadratojugal

More than 12 quadratojugals were collected, but the most complete detached specimen is a left quadratojugal (TMP 1988.55.144; Fig. 26) from a probable adult individual. It is 171 mm high and has a greatest mediolateral width at the back of the jugal suture of about 46 mm, the same as in a *P. canadensis* specimen (CMN 10646) from

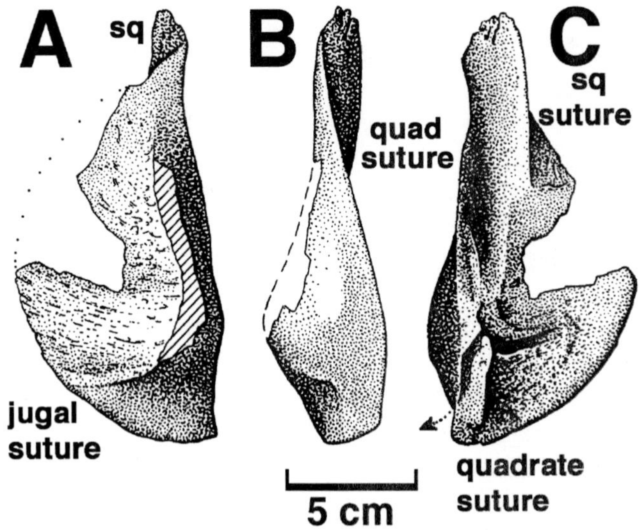

Fig. 26. Pachyrhinosaurus lakustai, *left quadratojugal (TMP 1988.55.114) in lateral (A), posterior (B), and medial (C) views. Arrow indicates course of the quadratic canal.*

Scabby Butte. It is a thin, wedge-shaped bone having only limited posteroventral exposure. In lateral view, it appears falcate with a slightly sigmoid posterior edge. It had a broad squamous contact with the jugal and a complicated contact with the quadrate. In *P. canadensis*, the contact is continuous, and the two bones are not separated by a quadratic foramen (Langston 1975). Contact with the forwardly descending process of the squamosal was through a modified syndesmosis that is a long slender slip joint in which the bones were secured by interosseous ligaments. The long ascending process was inserted between the anteroventral process of the squamosal and the quadrate. As in *Achelousaurus*, *Centrosaurus*, *Einiosaurus*, and *Styracosaurus*, the quadratojugal is excluded from the lateral temporal fenestra by the squamosal and jugal. In contrast, the quadratojugal of basal ceratopsians (You and Dodson 2004) and *Triceratops* bounds the lateral temporal fenestra posteroventrally (Hatcher et al. 1907). The bone apparently never fused with the jugal. It probably never fused with the quadrate either, although the deep, partly gomphoid, suture for the quadrate would have permitted little movement between these bones. Much of the quadrate canal is bounded by the quadratojugal. It is a narrow, 3 mm wide channel that passes upward close to the posterior edge within the sutural surface for the quadrate (Fig. 26C). The passage disappears above into an excavation that lodged a V-shaped ridge from the quadrate when the bones were connected. Posteroventrally, on the exposed surface between the jugal and the quadrate is a thumbnail sized scar that probably marks the origin of a minor division of the M. depressor mandibulae.

Quadrate

Quadrates are not present in any of the skulls so far prepared, but two dozen isolated specimens have been recovered; more than 75% of these are from the left side (Table 5). A left quadrate (TMP 1987.55.100, Fig. 27), lacking only the thin bone of the pterygoid flange, is one of the best examples. This bone is about three-quarters of the size of a *P. canadensis* quadrate from Scabby Butte (CMN 9610), and the width across the distal condyles is 77% that of a fragmentary quadrate (TMP 1993.29.4) from the Drumheller skull. It is more

Fig. 27. Pachyrhinosaurus lakustai, *left quadrate (TMP 1987.55.100) in anterior (A), posterior (B), lateral (C), and distal (D) views. Broken lines indicate continuation of pterygoid ramus (not preserved in any specimen). Dotted arrow indicates possible entry of the quadratic canal.*

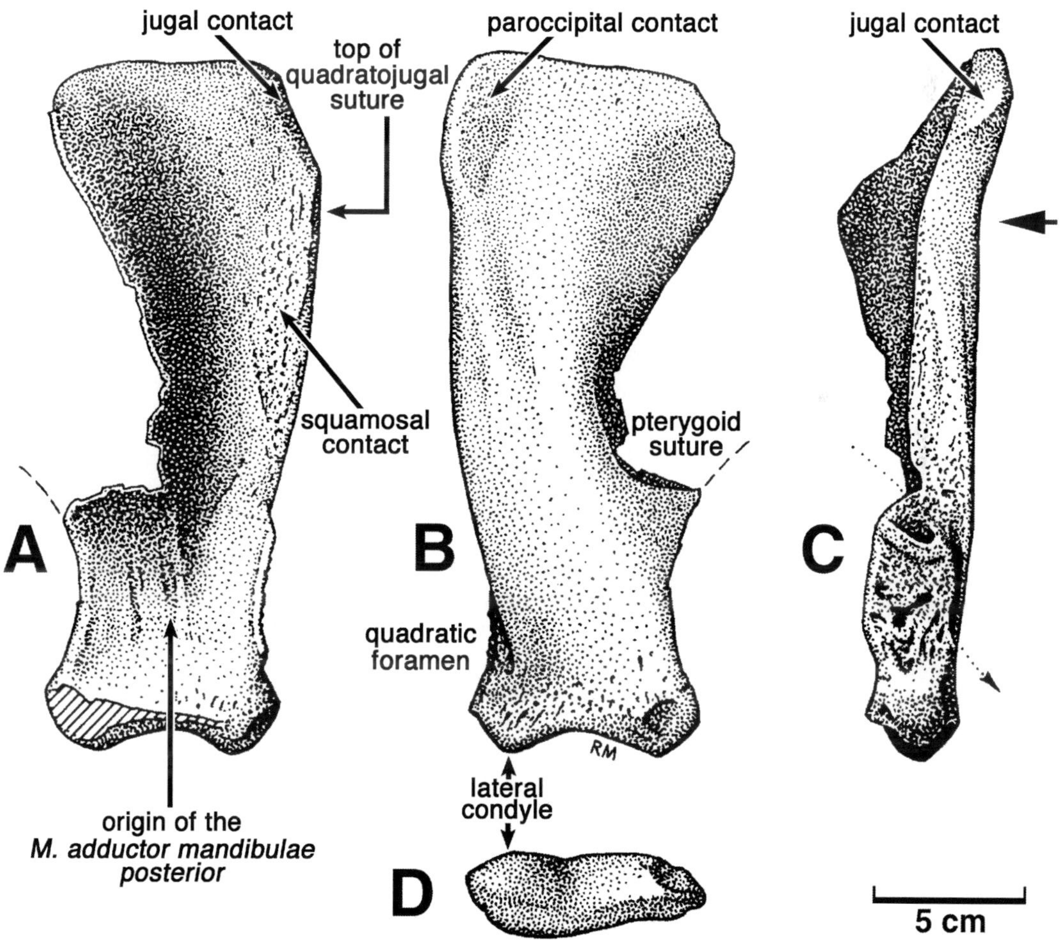

robust but is otherwise similar to the quadrate in *Centrosaurus*. When positioned in the skull, the thin obtuse proximal head of the quadrate was lodged in the narrow quadrate sulcus on the medial side of the squamosal. Viewed from behind, the lateral edge of the quadrate, which underlies the quadratojugal, is gently curved. Distolaterally, in this aspect, a small roughened triangular area above the lateral condyle indicates where the posteroventral edge of the overlapping quadratojugal curved around onto the posterior surface of the bone. There is a deep, dorsoventrally oriented groove on the posterior edge of the suture for the quadratojugal. A well-defined scar near the proximolateral corner of the posterior side marks the area overlapped by the lateral process of the exoccipital. This overlap was much more limited in extent than in *Triceratops* (Hatcher et al. 1907). The distomedial edge of the quadrate is broadly concave, continuing dorsally as the ventral edge of the expanded pterygoid flange (not preserved in available specimens). The depressed lateral edge of the flange area defines a wide, curved, and smooth face, sharp edged and deep below, but flattened out proximally, for contact with the quadrate ramus of the pterygoid. As in some specimens of *Centrosaurus*, the posteroventral corner of the surface extends into a pocket that enveloped the posteroventral corner of the quadrate flange of the pterygoid.

Seen from the side, the shaft of the quadrate is almost straight. Most of the lateral surface of the shaft is roughened, especially below, for attachment to the quadratojugal. An anteriorly directed flange, arising from the distal one-third of the bone, underlies the medial side of the quadratojugal. This flange, whose sutural surface is deeply scarred, is broader and shorter than the homologous, more acute, process in *Chasmosaurus*.

Two features are noteworthy on the anterior surface of the quadrate (Fig. 27A). Anterolaterally, at about midheight, is a long, smooth to slightly roughened surface that contacted the medial side of the quadrate ramus of the squamosal. It is roughly triangular, with its ventrolateral edge thickened and slightly raised. Laterally, the roughened surface bends sharply over onto the lateral side where it was attached to the quadratojugal. Dorsolaterally, the anterior and lateral corners of the quadrate are deflected forming a small tab that was wedged between the quadrate process of the jugal and the distolateral corner of the exoccipital. The second notable feature is the origin of the tendon for the *M. adductor mandibulae posterior* (Haas 1955; Ostrom 1966). It is a well-marked semicircular sulcate excavation about 38 mm tall and 19 mm wide, low down in the medial half of the anterior surface.

Table 5. Height of quadrate and distal width of condyles for *Pachyrhinosaurus lakustai* and *Pachyrhinosaurus canadensis* (TMP 1993.29.4).

Specimen No.	Side of skull	Height (cm)	Width (cm)*
TMP 1983.55.59	Left	—	88.0
TMP 1985.112.64	Left	—	83.0
TMP 1986.55.4	Right	—	33.0
TMP 1986.55.66	Left	—	45.0
TMP 1986.55.159	Left	—	70.0cr
TMP 1986.55.227	Left	—	72.5
TMP 1986.55.251	Left	—	72.0
TMP 1987.55.69	Right	—	60.0
TMP 1987.55.77	?	—	83.5
TMP 1987.55.100	Left	240	84.0
TMP 1987.55.101	Left	—	92.0p
TMP 1987.55.125	Left	—	73.0e
TMP 1987.55.146	Left	—	43.0
TMP 1987.55.180	Left	—	64.0
TMP 1987.55.191	Right	136	42.0
TMP 1987.55.221	Left	280	77.5
TMP 1987.55.292	Left	—	65.0
TMP 1988.55.41	Right	265	—
TMP 1988.55.23	Left	—	44.5
TMP 1989.55.107	Left	240	73.5
TMP 1989.55.145	Left	—	41.5
TMP 1989.55.337	Left	—	84.0
TMP 1989.55.615	Right	—	89.0
TMP 1989.55.1072	Right	—	79.0p
TMP 1993.29.4	Right	—	110.0

*Abbreviations: cr, crushing compromises accuracy of measurement; e, estimate; p, pathologic.

The quadratic canal was not identified with certainty on the quadrate of *P. canadensis* (Langston 1975). Its course passes mainly in the quadratojugals in centrosaurines, although a deep canal marks the anterolateral margin of the lower part of the quadratojugal suture and, presumably, marks the posteroventral course of the vein (Romer 1956). There is no obvious exit on the posterolateral edge of the bone in TMP 1987.55.100, but there is a shallow, 2 mm wide notch in TMP 1989.55.337 just above the posterior suture overlapped by a lappet of the quadratojugal. It may be that the 2 mm wide canal is simply obscured in most specimens by postmortem distortion.

The quadrate condyles (Fig. 27D) appear somewhat reduced anteroposteriorly compared with those of *P. canadensis*, but this may be a result of crushing.

Parietosquamosal frill

No intact parietosquamosal frill has yet been found for *Pachyrhinosaurus*. Even in the almost complete skulls, the fragility of the parietal bar and the lack of fusion between parietal and squamosal resulted in preburial separation of the back of the frill. Reconstructions of the frill of *P. canadensis* were based on disassociated pieces of parietals and incomplete squamosals (Sternberg 1950; Langston 1967, 1968, 1975). *Pachyrhinosaurus lakustai* frill material from Pipestone Creek includes numerous posterior segments of median parietal bars attached to incomplete posterior and lateral rami, a number of detached lateral rami, and many proximal sections of the median parietal bar. Several skulls preserve squamosals in position; TMP 1986.55.111 (Fig. 8E), TMP 1989.55.188 (Fig. 8B) and TMP 2002.29.1 have one squamosal apiece, and TMP 1989.55.1234 (Fig. 8F) retains both squamosals. Numerous disarticulated squamosals are also available. The many pieces of parietals from Pipestone Creek permit the reconstruction of the parietal part of the frill of *P. lakustai* with considerable confidence (Fig. 6). In plan view (as reconstructed), the frill is relatively narrower posteriorly and wider anteriorly than in other centrosaurines, and the lateral ramus of the parietal was probably relatively longer.

Frill ornamentation follows the scheme proposed by Sampson (1995). Process P2 is adjacent to the midline at loci 2, and the enlarged process P3 is the prominent anterolaterally curving hook. Fragments of the lateral parietal rami suggest that there were at least three more loci, possibly four, anterior to process P3. This suggests that there were seven marginal processes on each side of the parietals as in *Centrosaurus* (CMN 971), *Einiosaurus* (MOR 456 8-9-6-1), and *Styracosaurus* (CMN 344).

Squamosal

As in other centrosaurines, the squamosal of *P. lakustai* (Fig. 28) is short relative to skull length when compared with the condition in chasmosaurines (Lehman 1990). Its contribution to the frill is also limited and is relatively unadorned in comparison with the parietal part of the frill. It is impossible to determine the degree of sideward flaring of the frill, because all specimens have been flattened postmortem, and it is not possible to rearticulate individual bones to the skulls. Furthermore, with the exception of TMP 1989.55.188 (Fig. 8B) and TMP 1989.55.1234 (Fig. 8F), most of those still attached to skulls are incomplete. The Pipestone Creek collection contains more or less complete squamosals from juvenile to adult individuals. Among the best preserved are a left squamosal from a subadult (TMP 1989.55.268,

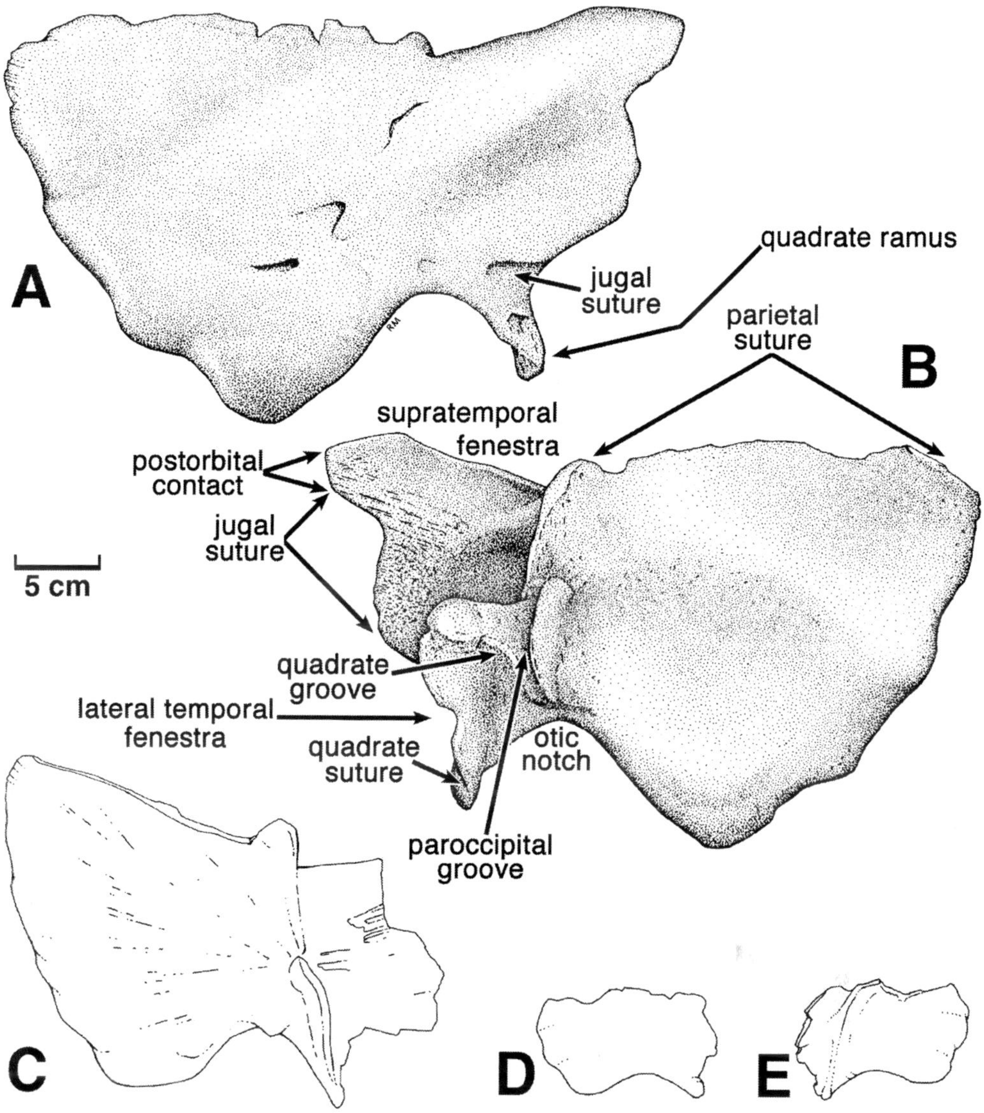

Fig. 28. Pachyrhinosaurus lakustai *squamosals. (A) Adult individual, TMP 1987.55.261, right lateral view; (B) the same, medial aspect; (C) TMP 1988.55.268, medial view; (D and E) TMP 1987.55.162 in lateral and medial views.*

Fig. 28C) and a large adult right squamosal (TMP 1987.55.261, Figs. 28A, 28B). The bone resembles closely those of *Einiosaurus* (Sampson 1993) and *Centrosaurus* but is relatively shorter than the latter. As in all other centrosaurines, the squamosal is conservative, showing little variability individually or ontogenetically (Fig. 28). Laterally, the squamosal generally has low broad marginal scallops. It differs from *Centrosaurus* mainly in the usual presence of only four such marginal scallops instead of five. The number varies somewhat in both taxa: three to six in *Centrosaurus* and three or four in *Einiosaurus* (Sampson 1993). The scalloped edge appears a little thinner than in *Centrosaurus* of similar size, and the scallops seem more regularly tabular. When viewed edge-on, the scallops appear as an undulating or wavy edge because the flanges are oriented in the same manner as those on the lateral ramus of the parietal (Sampson et al. 1997). However, those of the squamosal are more pronounced and regular in appearance. The dorsolateral surface of the squamosal is slightly arched diagonally from the edge of the supratemporal fenestra. In larger individuals, there are one or two low hummocks on this arch. Such elevations are common on large neoceratopsian squamosals, *Centrosaurus* and *Chasmosaurus* usually displaying two or more. The dorsal margin of the squamosal is "stepped-up" (sensu Dodson 1986) behind the dorsotemporal fenestra as in all other centrosaurines other than *Avaceratops*. The outer surface of the bone is lightly sculptured, and there are a few faint vascular sulci on the arched area and the hummocks.

As in other ceratopsids, the relationships of the squamosal to adjacent bones were loose and permitted growth throughout life.

What may be termed the otic recess (jugal notch of some authors) is almost identical in proportions and shape to that in a like-sized *Centrosaurus*. However, the anteroventral process that separates the notch from the lateral temporal fenestra is broader in *P. lakustai* than in *Centrosaurus*, and this fenestra was small, even for a ceratopsid.

The squamosal overlapped the postorbital anterodorsally and the jugal anterolaterally in squamous relationships. Although this arrangement appears to provide weak connections, the overlapping surfaces are broad and the squamosal–jugal contact was partly locked above the lateral temporal fenestra, where a small, posteriorly directed flange of the jugal passed over onto the lateral surface of the squamosal, producing a tongue-and-groove articulation.

The shallow tongue-and-groove union between squamosal and parietal is about one-half the length of the squamosal, which is the shortest in any ceratopsid. It would seem to have permitted some flexibility under stress, but the various ridges and grooves on the medial side for the quadrate and the paroccipital process are more strongly developed in TMP 1987.55.261 (Fig. 28B) than in a *Centrosaurus* squamosal of similar size. In particular, the area of the squamosal that contacted the quadrate is relatively more extensive in *P. lakustai*, indicating a more powerful suspensorium.

Parietal

The relationships between the frill bones and contiguous skull elements in *P. lakustai* conform to the pattern of all centrosaurines (Lehman 1990). The parietal seems less robust than that in *Centrosaurus*, perhaps owing to loss of an unknown amount of the thin edges bordering the parietal fenestrae. The frill is fenestrated, but the sizes of the openings are unknown. Tapered thinning of the edges of parietal bars, and the posterior and lateral parietal rami would suggest fontanelles roughly comparable in size with those in *Centrosaurus*. Around the margins of the frill is a bizarre array of variable bony spikes, hooks, horns, and marginal flanges (Fig. 6).

In *P. canadensis* (CMN 8867 and the Drumheller skull) where the anterior parts of the parietals are in situ, the superior profile of the parietal bar is approximately parallel to the frontal plane of the skull. This also seems to have been true in TMP 1986.55.258 (Fig. 7). However, several incomplete posterior parietal bars of *P. lakustai* curve upward posteriorly in broad arcs. In a subadult, the arc subtends approximately 50°. Thus, our reconstruction of the frill (Fig. 6B) is elevated posteriorly on the order of the frill in *Centrosaurus apertus* (AMNH 5239, Brown 1914*b*). The shortness of the squamosal in *P. lakustai* is consistent with this interpretation.

The median parietal bar is variably triangular in section anteriorly (Figs. 29C, 29D, 30C, 30F, 30I, 30J). Proximally, beneath the parietal horns, its ventral side is deeply excavated or vaulted (Figs. 30A, 30C, 30F, 30I, 30J, 30L, 30M), differing substantially from the broadly convex cross sections in other large ceratopsians. This architecture is no doubt a device for strengthening the relatively slender design of the frill with its array of massive spikes. The bar flattens dorsoventrally towards the back, where the vaulting largely disappears, being replaced gradually by a flat or even slightly downwardly convex surface.

The median bars of smaller parietals of *P. lakustai* have sagittally varied series of laterally compressed, broadly rounded triangular humps (Fig. 29). In larger examples, these humps impart a longitudinally serrated dorsal profile to each parietal bar. Because the leading edges of the humps are longer than the trailing edges, they often create a somewhat backwardly raked comb-like appearance. Similar humps, termed low longitudinal swellings by Gilmore (1917) and low, round, rugose, median prominences by Hatcher et al. (1907), are common in centrosaurines (Sampson 1995), and there can be a series of as many as eight in *Centrosaurus* (Ryan 2003). They are most strongly expressed in *P. lakustai*. The number of humps varies, and the most anterior hump, or the one following it, is usually the largest of the series.

Surely, the most surprising feature of the *P. lakustai* frill is the presence in all adult individuals of well-developed median horns rising anterodorsally from the parietal bar. The holotype is the best preserved example because the horn-bearing part of the parietal bar is attached to the skull (Figs. 7A, 30A, 30B). This horn is slightly curved, and its height is about 1.7 times its basal length. As preserved, it is only about one-half as wide as its basal length, but it appears to have been somewhat compressed laterally by crushing. Its tip is blunt, perhaps owing to abrasion, whereas some other horns are more pointed. All examples of parietal horns are trapezoidal in basal plan, their flattened faces joining at obtusely rounded edges (Fig. 30K). The anterior side is generally about twice as wide as the posterior side. All horns have a radial constriction around the base (Figs. 30A, 30B, 30E, 30F,

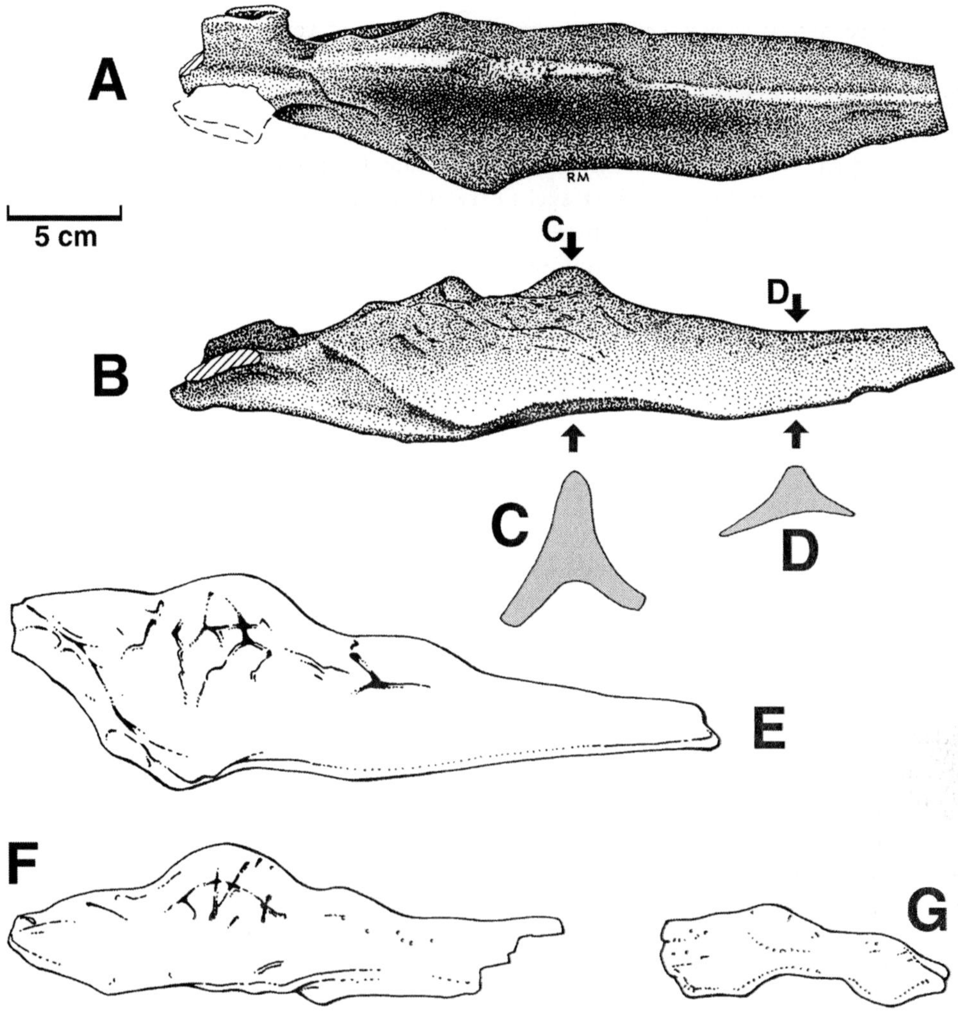

Fig. 29. Pachyrhinosaurus lakustai, *anterior segments of median parietal bars (anterior to left), showing sagittal humps. TMP 1987.55.252 in dorsal (A), lateral (B), and cross section (C and D). Graded series illustrating development of the comb-like appearance of the median parietal humps in TMP 1986.55.127 (E, drawn from the right side, then image reversed), TMP 1985.55.47 (F, reversed) and TMP 1987.55.187 (G).*

30H, 30J, 30K, 30L) reminiscent of the burr on the horns of some ruminants. Also, like mammalian horn cores, there is a series of nutrient foramina that enter the core in the basal constriction, and coarse vertical striations adorn the external surfaces. Similar to some bovine horn cores, the fluted appearance on most of the horns is partly due to subparallel superficial vascular channels passing upward from the base. However, other more regular parallel channels appear to be internal tubular canals accidentally exposed postmortem. The pithy structure suggests copious blood supply and the presence of some sort of thick integumentary covering or sheath. The parietal horn in TMP 1986.55.258 (Fig. 30B) is 200 mm high and 120 mm long at the base but is surpassed in size by several incomplete examples in the collection. The center of its base is positioned about 170 mm behind the area of the transverse frontoparietal contact, which is ill defined in the specimen.

Several individuals each have a single parietal horn. However, in TMP 1983.55.24, there is a small nub just in front of the horn (Fig. 30E). It is horn-like in texture and structure, but its top is obtuse and hummocky. An even more extreme development of ornamentations occurs on a short piece of a large parietal bar (TMP 1988.55.187, Figs. 30D, 30K). Immediately in front of the base of the principal horn is a forwardly curved, smaller horn. The anterodorsal face of this curved horn has a rounded rough-bottomed facet (Figs. 30D, 30K, 30M), possibly the site of a detached osteoderm. Behind the principal horn, there is also a nub of bone (Figs. 30J, 30K) resembling that of

Fig. 30. Pachyrhinosaurus lakustai, *median parietal horns. Holotype TMP 1986.55.258 seen from the left side (B) and in transverse sections (A and C); (D) TMP 1988.55.187 seen from above; (E) TMP 1983.55.24 seen from the left side; (H) TMP 1986.55.211 seen from the left side; transverse sections (F and I) and horizontal section (G) of horn of TMP 1986.55.211; (K) TMP 1988.55.187 seen from the right side; (J, L, and M) transverse sections of Fig. 30K.*

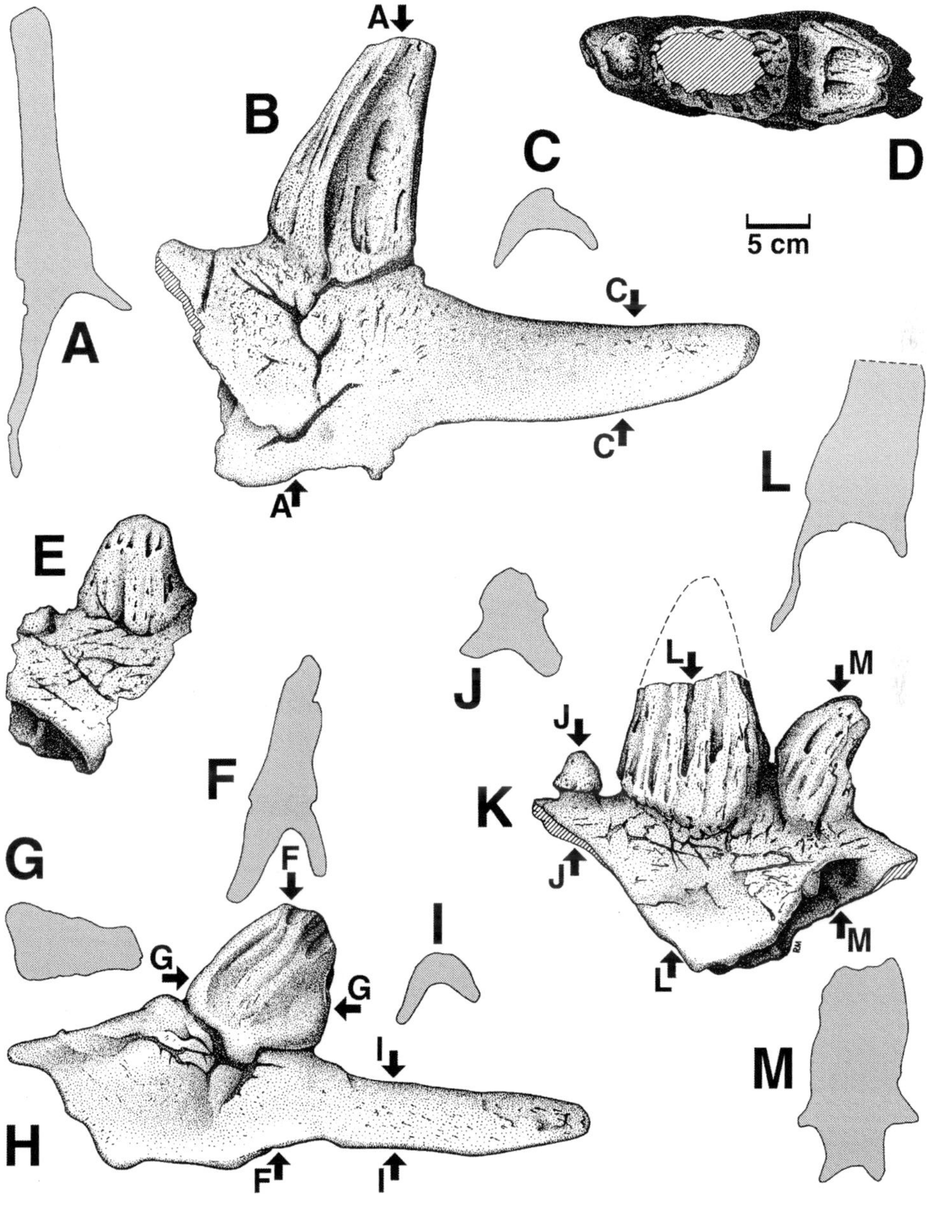

TMP 1983.55.24 (Fig. 30E). It has a deep annular groove at its base, suggesting fusion of an osteoderm to the parietal. Indeed, an isolated osteoderm, TMP 1987.55.323 (Fig. 31) resembles this nub, although it could also be an epijugal, epoccipital, or one of the ornamentations of the promaxillary comb. If the nub on TMP 1988.55.187 (Fig. 30K) is an osteoderm recently fused to the parietal, that in front of the horn in TMP 1983.55.24 (Fig. 30E) might represent a somewhat more advanced ontogenetic stage of osteodermal fusion. Thus, all parietal horns may be osteoderm derivatives subject to a high degree of variability. However, in one example (TMP 1988.55.26), a low blunt parietal horn appears to be simply an upgrowth of a median hump, and there is no indication that it was capped by an osteoderm.

Fig. 31. Osteoderm or epijugal associated with Pachyrhinosaurus lakustai *remains in the Pipestone Creek bone bed. TMP 1987.55.323 in dorsal (A) and side (B) views. The anatomical position of this bone is unknown.*

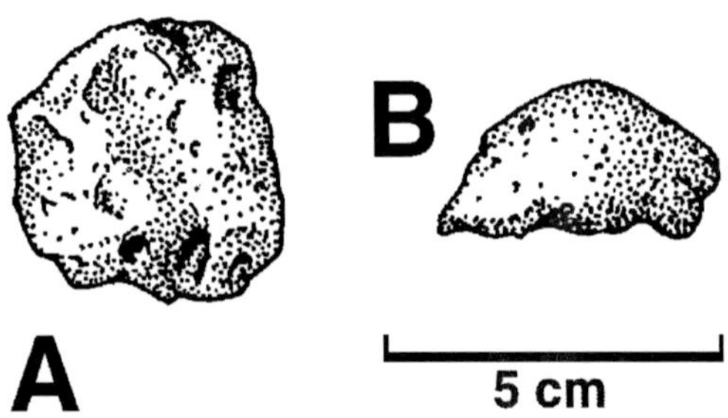

The number of adventitious horns might have increased with age, but there are several massive parietals that have but a single horn each and lack any sign of nubs. Some of these horns are lower and blunter than those just described, but the unaccompanied horn of the holotype is amongst the larger examples. The numbers and sizes of the parietal horns are variable features in adult animals. Most commonly, there are only single horns, but occasionally there are three. Where a parietal horn is present, humps are insignificant or lacking. Moreover, horn-bearing parietals are always larger than the hump-bearing specimens, suggesting that the humps became incorporated into the horns. Alternatively, the humps may have disappeared during ontogenetic thickening of the parietal bar. In conclusion, it is possible that parietal humps ontogenetically preceded the incorporation of osteoderms, which later in life developed into parietal horns in at least one of the sexes.

The posterior edge of the parietal frill is shallowly emarginated. Unlike *Centrosaurus* and *Styracosaurus*, *P. lakustai* lacks the hook-like process 1 near the midline (Sampson 1995). *Achelousaurus*, *Einiosaurus*, and *P. canadensis* apparently also lack process P1 (Sampson et al. 1997). On either side of the emargination (which varies considerably in width and depth), a medially hooked, horn-like process (process P2 of Sampson 1995; Sampson et al. 1997) arises from the posterior edge in many of the larger frills. One or both of these hooks may cross the sagittal plane and overlap its opposite number—one growing above the other, thus avoiding direct contact (Fig. 32C). Some frills have only one hook on each side (Fig. 33I), others have none (Fig. 33C), and several have only small elevations in place of hooks (Figs. 33B, 33D). Some of this numerical variation is probably related to predepositional weathering. Some of the smallest hooks occur on the most massive parietals (Fig. 32A). In some frills, the hooks are flattened dorsoventrally, whereas they are round in section in others (usually larger ones) and resemble short bananas. In TMP 1987.55.210 (Fig. 32C), the right hook was at least 150 mm long; it crosses the sagittal plane a little beyond its midlength. The opposite hook is smaller and more tightly curved, crossing the midline at about two-thirds of its length and passing beneath the larger hook at a distance of about 20 mm. This crossing of hooks creates a semicircular opening visible when the frill is viewed from the front (Fig. 32C). An annular groove marks the base of each hook ventromedially but is scarcely discernible elsewhere. As this basal groove passes

Fig. 32. Pachyrhinosaurus lakustai *incomplete parietal frills in dorsal plan view. Specimens were selected to show variability in the expression of spikes and hooks. (A) TMP 1987.55.258; (B) TMP 1987.55.232; (C) TMP 1987.55.210; (D) TMP 1987.55.141; (E) TMP 1989.55.258.*

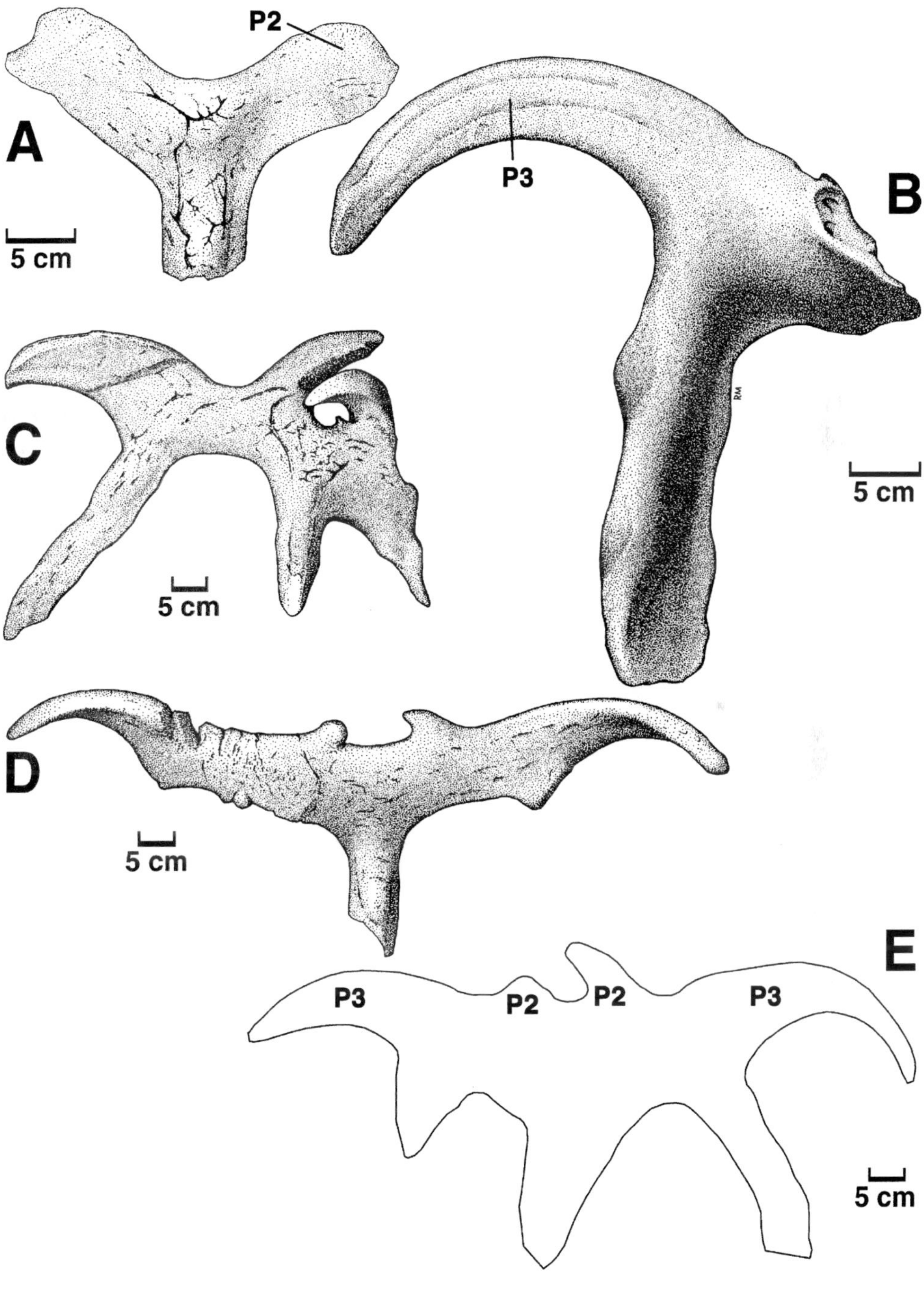

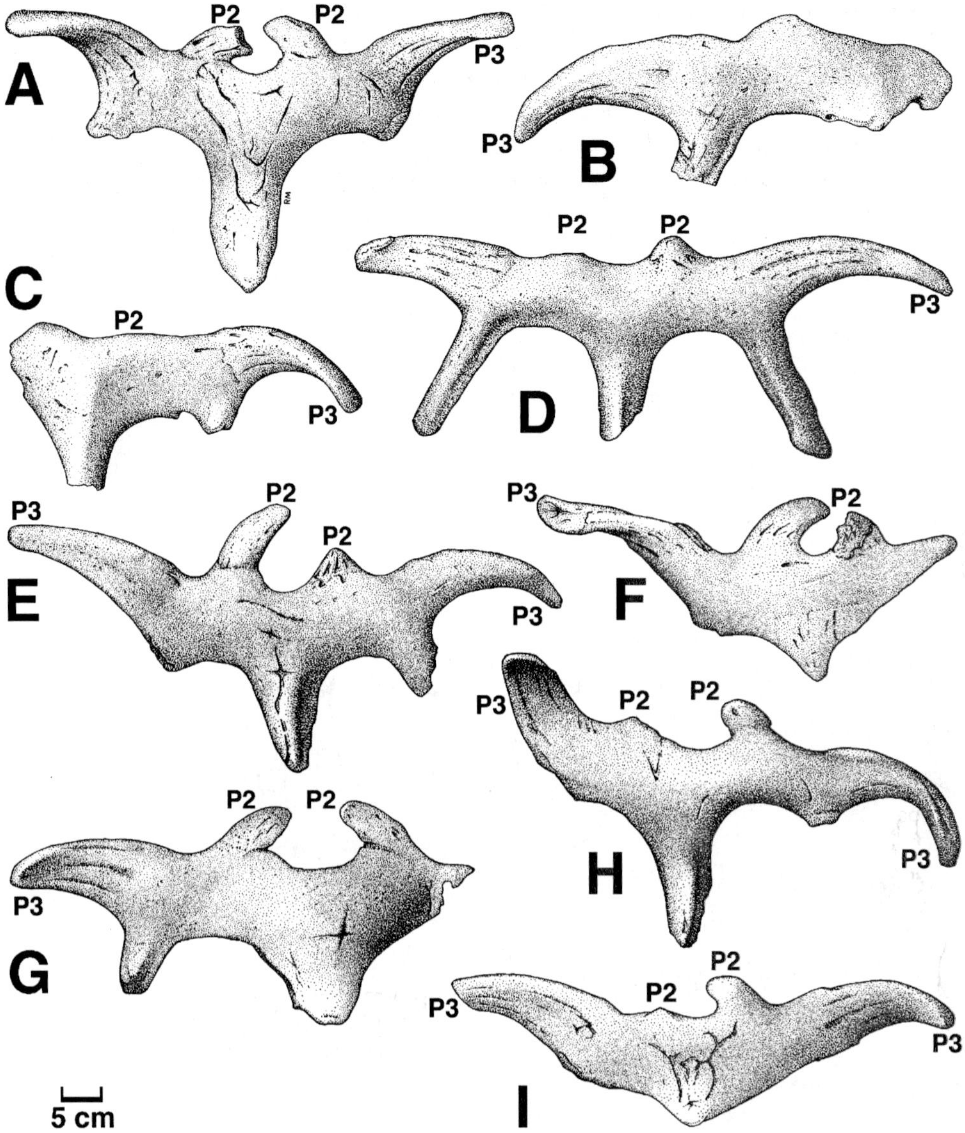

Fig. 33. Pachyrhinosaurus lakustai *parietals in dorsal (plan) view: (A) TMP 1989.55.1241; (B) TMP 1987.55.164; (C) TMP 1986.55.113; (D) TMP 1989.55.1144; (E) TMP 1989.55.1503; (F) TMP 1986.55.239; (G) TMP 1986.55.261; (H) TMP 1989.55.1085; (I) TMP 1988.55.46.*

posteromedially and diagonally across the lower part of the hook, it deepens into a fissure that separates a small "secondary" horn-like process at the medial base of each hook (Fig. 32C). We have not observed this condition in other ceratopsians; at first, we regarded it as a postmortem effect. However, it occurs on several hook-bearing frills where there is no evident damage. We offer no suggestions as to its functional significance.

A massive forward-curving cornuate spike (process P3 of Sampson 1995) originates on the posterolateral edge of the posterior ramus of the parietal. Such spikes, like the other ornamentations on the frill, are highly variable and occasionally asymmetrical. Most of them are somewhat twisted distally with torsion amounting to as much as 140° in some specimens (Figs. 32D, 33A, 33F). In one relatively symmetrical frill (TMP 1987.55.141; Fig. 32D), the lateral spikes span a distance of 1.15 m. Two or three strong subparallel longitudinal grooves appear distally on the spike, and the combination of all these features is reminiscent of some bovine horns. Although the grooves are variable, one that passes along the posterodorsal surface from near the base of the spike seems more constant in position than the others. The spikes generally have flattened ovate cross sections (sometimes accentuated by crushing). Although it is unclear whether the hooks and spikes on the frills are of osteodermal origin or are simple outgrowths of the parietal bone, it seems probable that they are specialized epoccipitals (Sampson et al. 1997).

A parietal spike of *P. canadensis* from Scabby Butte is straighter than any of the spikes from Pipestone Creek and is not twisted (Langston 1975). A puzzling horn-like object, also from Scabby Butte (Langston 1975), may be a weathered parietal spike that is about one-third larger than any specimen from Pipestone Creek.

Marginal scallops (processes P4–P7 of Sampson 1995) are generally smaller (Fig. 34) than in *Centrosaurus*. Five such scallops are present on the most complete examples, and this is likely the usual number. In contrast, the number of scallops displayed in other centrosaurine parietals is usually four or five in *Centrosaurus*, four in *Einiosaurus*, three in *Achelousaurus*, and two in *Styracosaurus*. Each scallop consists of a thick, broadly curved lateral flange bounded anteriorly and posteriorly by a shallow emargination in the edge of the parietal. The row of scallops crosses the edge of the parietal diagonally so that the posterior end of each flange is positioned higher on the frill than its anterior end (Fig. 34F; Sampson et al. 1997). Orientations of the imbricated flanges vary in *Centrosaurus* and appear eccentric in some specimens (just as in *P. lakustai*) or lie more in the plane of the frill breadth. Usually the largest scallop in the frill is positioned posteriorly just anterior to the base of the P3 parietal spike (Fig. 34A), but this scallop is not developed in some examples. The surface textures of the scallops are coarser than that of the parietal bar, probably signaling differences in the overlying integument. In one specimen (TMP 1989.55.1030), three distinct ossicles are partly coalesced, cap-like, to the underlying flanges (Figs. 34F, 34G).

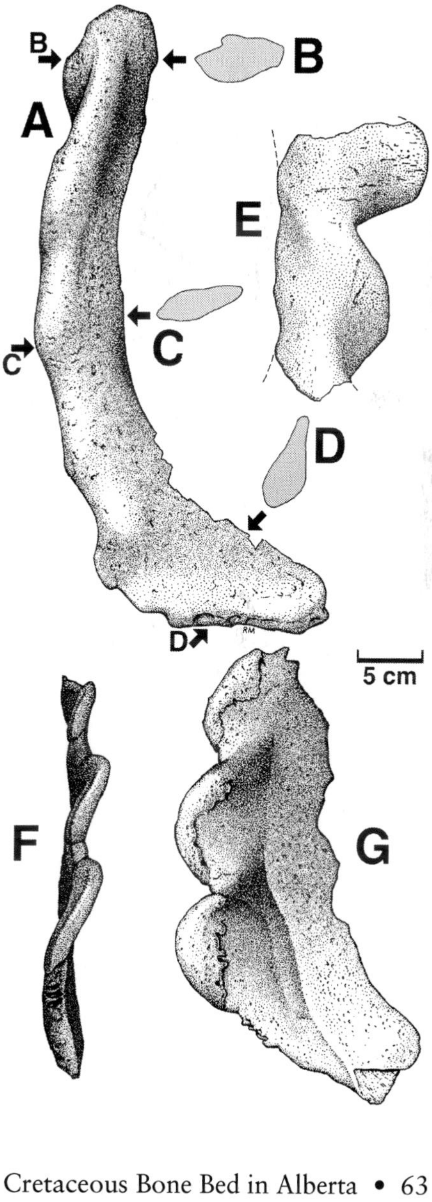

Fig. 34. Pachyrhinosaurus lakustai, *incomplete lateral rami of parietals: (A–D) TMP 1986.55.193, dorsal plan view (A) and cross sections (B, C, and D); (E) dorsal view of TMP 1983.55.15; (F) lateral view of three most proximal lateral flanges of TMP 1989.55.1030 (posterior is to the left; note how the flanges rise posteriorly, crossing the transverse plane of the parietal); and (G) TMP 1989.55.1030 in dorsal aspect.*

Large parietal frills (TMP 1986.55.157, Fig. 35A; TMP 1989.55.757, Fig. 35E) and fragments of several slightly smaller ones (Figs. 35B–35D) differ significantly from those just described. TMP 1986.55.157 (Fig. 35A) is about the same size as the large parietals with spikes and hooks but displays neither of these features. The bone is much thinner and more plate-like, is more deeply emarginated posteromedially, and has broad flat scallops on its posterior edge. More of the parietal bar is preserved than usual in this specimen and, anteriorly, it has four median elevations; the most forward one is the most strongly expressed. These frills are more suggestive of "*Monoclonius*" than of pachyrhinosaurs but probably represent an ontogenetic stage (Sampson et al. 1997).

Fig. 35. Pachyrhinosaurus lakustai, *incomplete "Monoclonius"-like parietals of immature individuals in dorsal views.*
(A) TMP 1986.55.157;
(B) TMP 1989.55.125;
(C) TMP 1989.55.499;
(D) TMP 1987.55.258;
(E) TMP 1989.55.757.

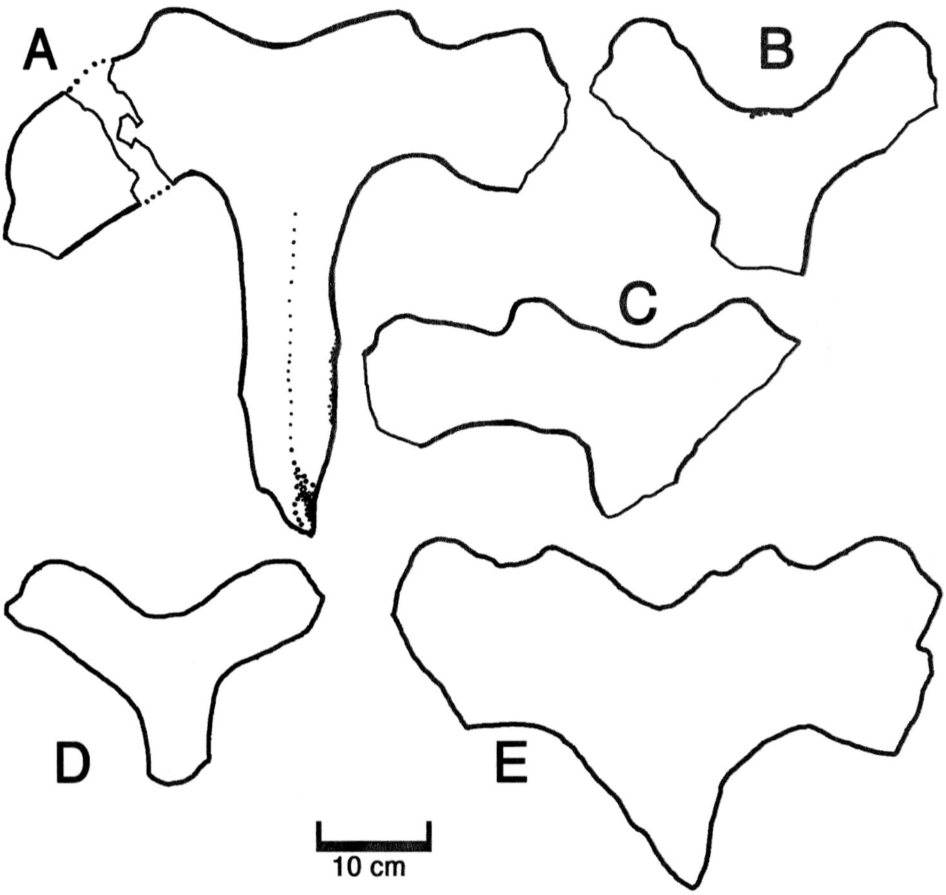

The frill of the closely related *Styracosaurus albertensis* has often been illustrated but never with the obvious dorsoventral distortion corrected. However, it may have resembled *P. lakustai* and *P. canadensis* in its orientation. Such an erect and ornamented frill would have presented an imposing display in head-on confrontation and would conform to current ideas about the functions of ceratopsian frills (Farlow and Dodson 1975; Kurzanov 1972; Spassov 1979; Sampson 1997*b*).

Epoccipitals

Twenty unfused epoccipitals were recovered from the Pipestone Creek bone bed. From their sizes, it is evident that they ossified relatively late in ontogeny and co-ossified shortly after to the frill elements. Such unfused epoccipital ossifications are sometimes found in *Centrosaurus* but are not common. The smallest well-preserved epoccipital (TMP 1989.55.931) from the Pipestone Creek bone bed is a small crescentic element 60 mm long, 20 mm proximodistally, and 11 mm thick. TMP 1989.55.566 is the largest free element and is 65 mm long, 62 mm wide and 27 mm thick. The proximal end is divided into two subequal sutural faces, as though it formed a wedge between two adjacent scallops. A pair of sutural facets is also present on TMP 1989.55.656, but one is almost 3 times the length of the other. This specimen is also thick (28 mm) with heavily ridged sutural surfaces, which suggests it was close to becoming co-ossifed at the time of death. The presence of shallow grooves around the bases of many of the scallops in mature skulls and the presence of rings of small foramina indicate that epoccipitals are fused to the margins of parietals and squamosals in all mature individuals.

Palate

The six isolated palatal elements that have been recognized are fragmentary (Fig. 36), and external cranial bones generally obscure those associated with skulls (Fig. 37). The left pterygoid, palatine, and ectopterygoid are partially visible in medial view from the right side of TMP 1989.55.188 (Fig. 37) because of the loss from the right side of the jugal and back of the maxilla and do not look significantly different from the same elements in *Triceratops* (Hatcher et al. 1907, Fig. 20). Each of two fragments of pterygoids (TMP 1989.55.249, TMP 1989.55.415) represent a section of the quadrate process and the region adjacent to the basipterygoid process (Fig. 36).

TMP 1986.55.312 is a pair of symmetrical fragments that we have not been able to identify, even though they are similar to the pterygoid fragments in texture, thickness, and general contours. Two other fragments have pedestals with distal facets that are scarred by conspicuous ridges and grooves. These may represent the sutural facets of the palatine for the pterygoid. Without adequate comparative material at hand, the identification remains tentative.

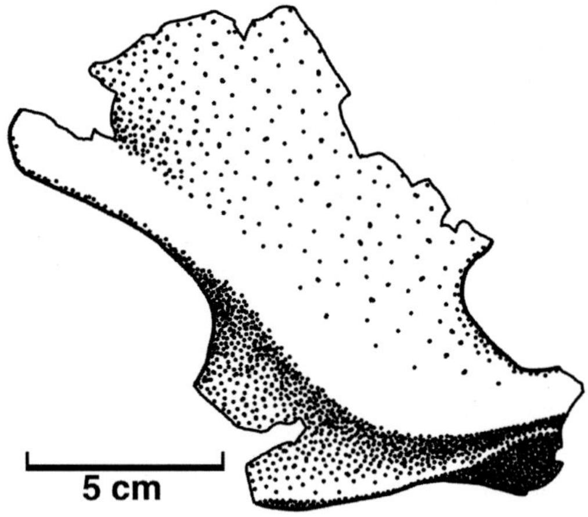

Fig. 36. Base of quadrate process of left pterygoid (TMP 1989.55.249) in medial view.

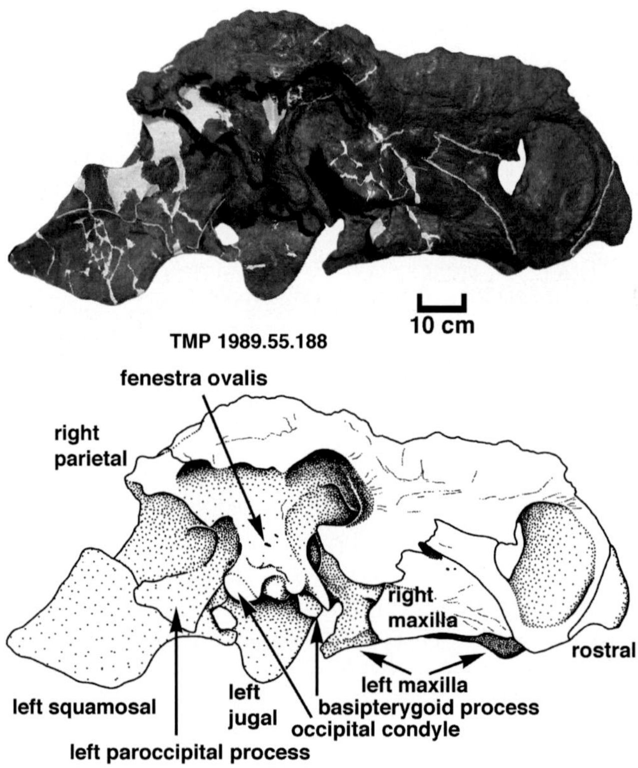

Fig. 37. Pachyrhinosaurus lakustai *(TMP 1989.55.188), right side of the skull showing braincase*

Neurocranium

The braincase is well represented in articulated skulls of *P. lakustai* (TMP 1986.55.111; TMP 1986.55.258, Fig. 7; TMP 1987.55.156, Fig. 8D; TMP 1987.55.304; TMP 1989.55.188, Figs. 37, 38; TMP 1989.55.1234), although they are invariably at least partially obscured by the outer skull bones or matrix. Two isolated braincases (TMP 1989.55.517, Fig. 39; TMP 1989.55.1243, Fig. 40) provide the best information on the cranial nerves.

Fig. 38. Pachyrhinosaurus lakustai *(TMP 1989.55.188), close-up of braincase posteroventral–lateral view.*

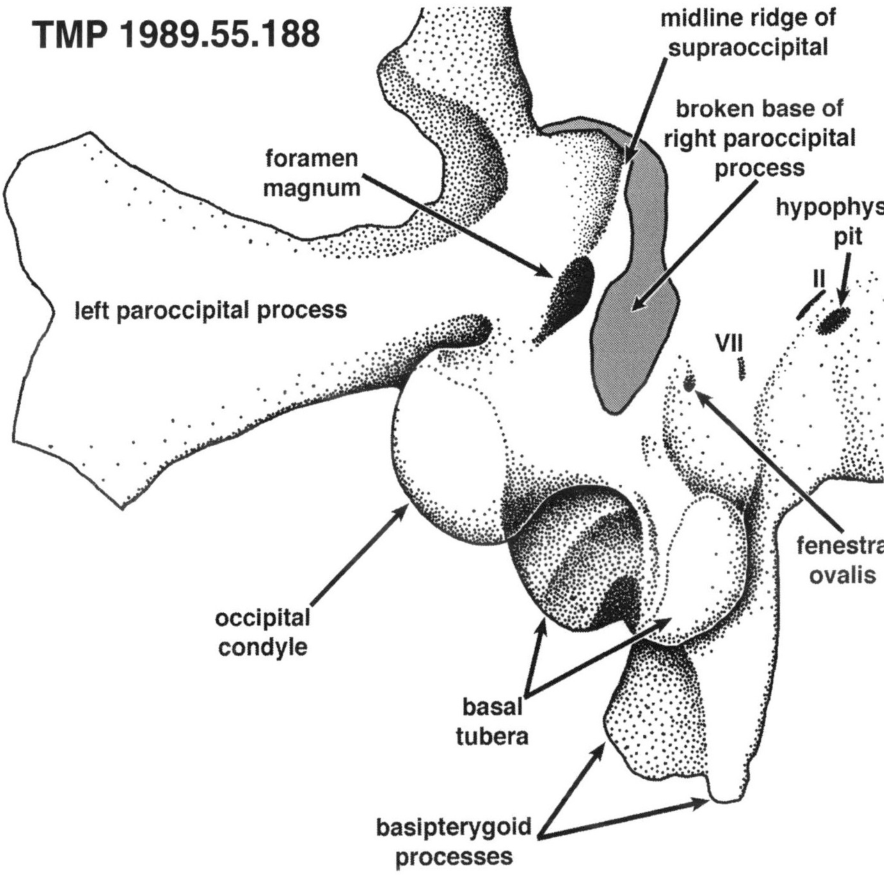

Ceratopsid braincases are distinctive in comparison with those of other vertebrates but are relatively conservative within the family, varying interspecifically mostly in proportions that are correlated with differences in overall cranial dimensions and morphology. Anatomical details of individual braincase bones are usually obscured by fusion with adjacent elements.

The articular surfaces of undistorted occipital condyles are almost round in *P. lakustai*, as in all other known ceratopsids (Dodson and Currie 1990). A bivariate analysis of width versus height in occipital condyles of 50 ceratopsids (data from Ryan 1992; Anderson

Fig. 39. Pachyrhinosaurus lakustai *braincase (TMP 1989.55.517), which is split horizontally into two sections. (A) Ventral view of dorsal section; (B) dorsal view of ventral section looking at the floor of the endocranial cavity. With the exception of the darkest gray shading of the roof of the endocranial cavity, gray shading represents broken bone surfaces. Roman numerals are the cranial nerve numbers. V¹, ophthalmic branch of cranial nerve V.*

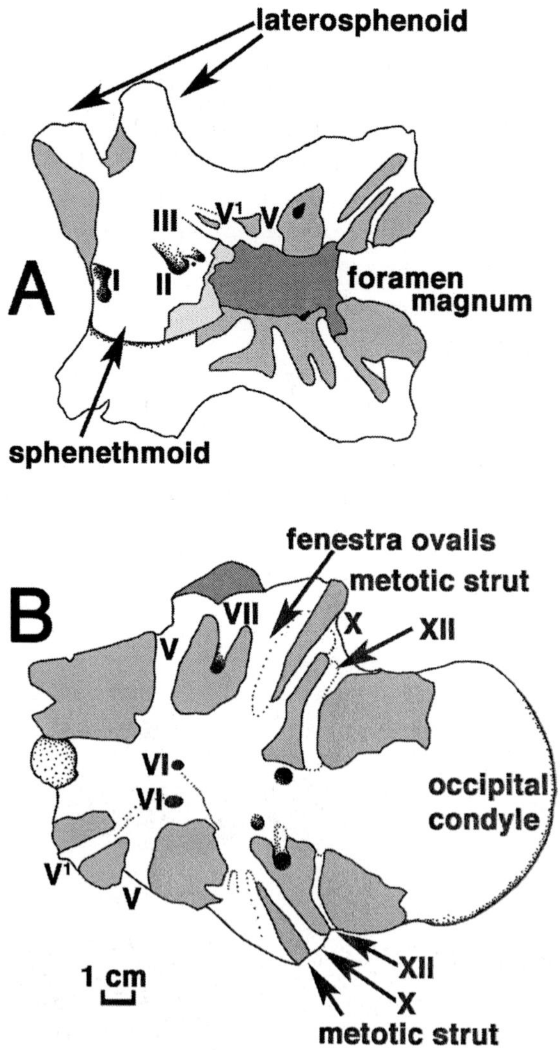

1999; Goodwin et al. 2006; and specimens in the Tyrrell Museum) shows that the relationship between these two dimensions remains the same throughout life (the coefficient of allometry is 0.9989). In several large detached occipital condyles of *P. lakustai*, the exoccipital and basioccipital components are completely fused into the characteristic ceratopsian condylar sphere. The condyles range in diameter from 68 to 72 mm (transverse), and from 69 to 78 mm (dorsoventral). As in articulated skulls of *P. canadensis* (Sternberg 1950), the condyle is oriented more ventrally than that of *Centrosaurus*. Even the largest condyle from Pipestone Creek is only about three-quarters of the size of the condyle in the Scabby Butte (CMN 9485 is 95 mm wide by 92 mm high) and Drumheller skulls (90 mm wide by 100 mm high, possibly distorted). Several subadult condyles (TMP 1985.112.66, TMP 1986.55.213, TMP 1987.55.133, TMP 1988.55.126,

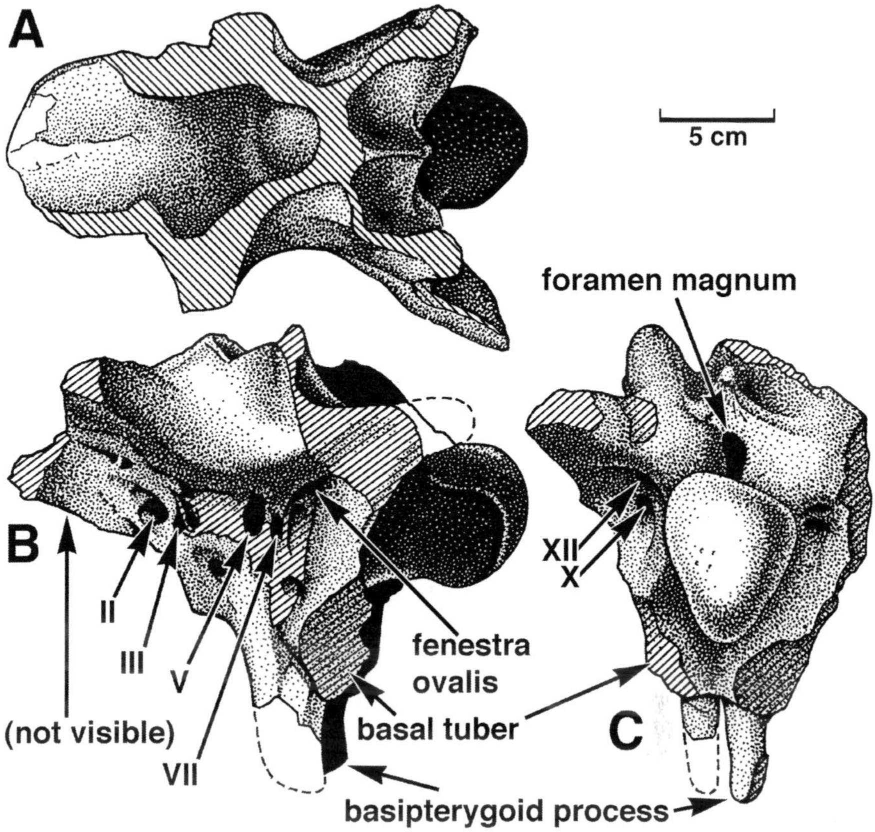

Fig. 40. Pachyrhinosaurus lakustai, *braincase. Most of an isolated braincase (TMP 1989.55.1243) in dorsal (A), lateral (B), and posterior (C) views.*

TMP 1988.55.237) show the characteristic ceratopsid tripartite division by the basioccipital and the two exoccipitals of the condyle. As in other ceratopsians, the basioccipital is excluded from the floor of the foramen magnum by the exoccipitals.

As expected for a large animal, the foramen magnum is small compared with the occipital condyle (Fig. 38). Because the spinal cord is negatively allometric during growth and the occipital condyle exhibits positive allometry, the size relationship between these two dimensions is size dependent. Differential allometry also probably accounts for the fact that the basal tubera are relatively smaller in relation to occipital condyle size in pachyrhinosaurs (Langston 1975).

As in most dinosaurs, the exoccipital and opisthotic fuse indistinguishably to form the lateral margins of the foramen magnum and the elongate paroccipital process (Figs. 37, 38). Additionally, the exoccipitals surround the foramen magnum, thus excluding the supraoccipital and basioccipital from the opening (Dodson et al. 2004). The left paroccipital process of TMP 1985.55.188 (Fig. 38) is 24.5 cm long and expands distally to a height of 14 cm. Two foramina exit the exoccipital at the base of the condylar neck (the higher, smaller one is for cranial nerve XII, whereas the larger lower opening subdivides internally for cranial nerves X and XI) in this specimen and in TMP 1989.55.1243 (Fig. 40C) as in *P. canadensis* (Langston 1975) and all other ceratopsids.

The supraoccipital is fused indistinguishably to the exoccipital–opisthotic complex in all specimens. There is the usual midline supraoccipital ridge (Figs. 38, 40A), separating the attachment surfaces of the M. rectus capitis (Dodson et al. 2004). Participation of the supraoccipital in the margin of the foramen magnum seems to be a size-related feature in ceratopsians (Horner and Goodwin 2006) but as noted, it is separated from this opening by the exoccipital–opisthotic complex in all mature ceratopsids, including those from the Pipestone Creek bone bed.

In all mature specimens where the anterior part of the braincase is preserved, the interorbital septum is ossified and fused to the cultriform process and ethmoid ossifications. It is also invariably mostly hidden by the outer bones of the skull (Figs. 37, 38), so no distinctive characters can be perceived. The opening for the optic nerve (cranial nerve II) is surrounded by the orbitosphenoid, which is indistinguishably fused to the laterosphenoid and the more anterodorsal sphenoid ossifications in all specimens where it is present.

The laterosphenoid is robust and thick, extending anterodorsally to meet the downturned triangular process of the frontal and postorbital. In dorsal view, there is a concave depression between the laterosphenoids that extends forward above the ethmoid ossifications and backward to the supraoccipital. This ossified floor of the frontal fontanelle is formed by the frontals, which in turn roofs the braincase (Sternberg 1927). Ventrolaterally, the laterosphenoid forms the anterior borders of the large foramen (fenestra ovale) for cranial nerve V. The ophthalmic branch of cranial nerve V separates within the laterosphenoid from the more posterior branches and exits anteriorly through a large foramen dorsolateral to the exit for cranial nerve VI.

An isolated basisphenoid (TMP 1988.55.126; Fig. 41) has a deep, posteriorly oriented suture with the basioccipital that is notched ventrally between the basal tubera. The basisphenoid forms the anterior margin of each basal tuber but bends sharply anteroventrally to form the basipterygoid process. The basal tubera and basipterygoid process are separated by a narrow notch in *P. lakustai*, whereas this concavity is more open in *Centrosaurus*. This correlates with the more downturned occipital condyle (Fig. 37) of pachyrhinosaurs. A canal extends dorsally from this notch on the lateral surface and enters another notch just below the triangular suture for the exoccipital–opisthotic, prootic, and possibly laterosphenoid. Dorsally, the basisphenoid formed the floor of the braincase for a short distance (17 mm). The pituitary entered a deep pit through the infundibular foramen, the back margin of which was formed by the basisphenoid (dorsum sellae). The cultriform process was not recovered with TMP 1988.55.126. There is a single small opening (Fig. 41D) on the back wall of the sella turcica that might have been for the common carotid as in *Triceratops* (Forster 1990), but two larger openings, one on each side of the midline (Fig. 41D), more likely represent the passages of the internal carotids. A foramen for the cranial nerve VI exits the anterior surface of the basisphenoid on each side of the sella turcica. The basipterygoid processes diverge ventrally at an angle of about 45° (Fig. 41A), whereas the basisphenoid contributions to the basal tubera are short and are separated by a shallow notch (Fig. 41B).

Fig. 41. Isolated basisphenoid (TMP 1988.55.126) in anterior (A), posterior (B), and lateral (C) views. (D) Anterodorsal view of pituitary fossa showing the paired openings for the internal carotids, and the more lateral exits for cranial nerve VI.

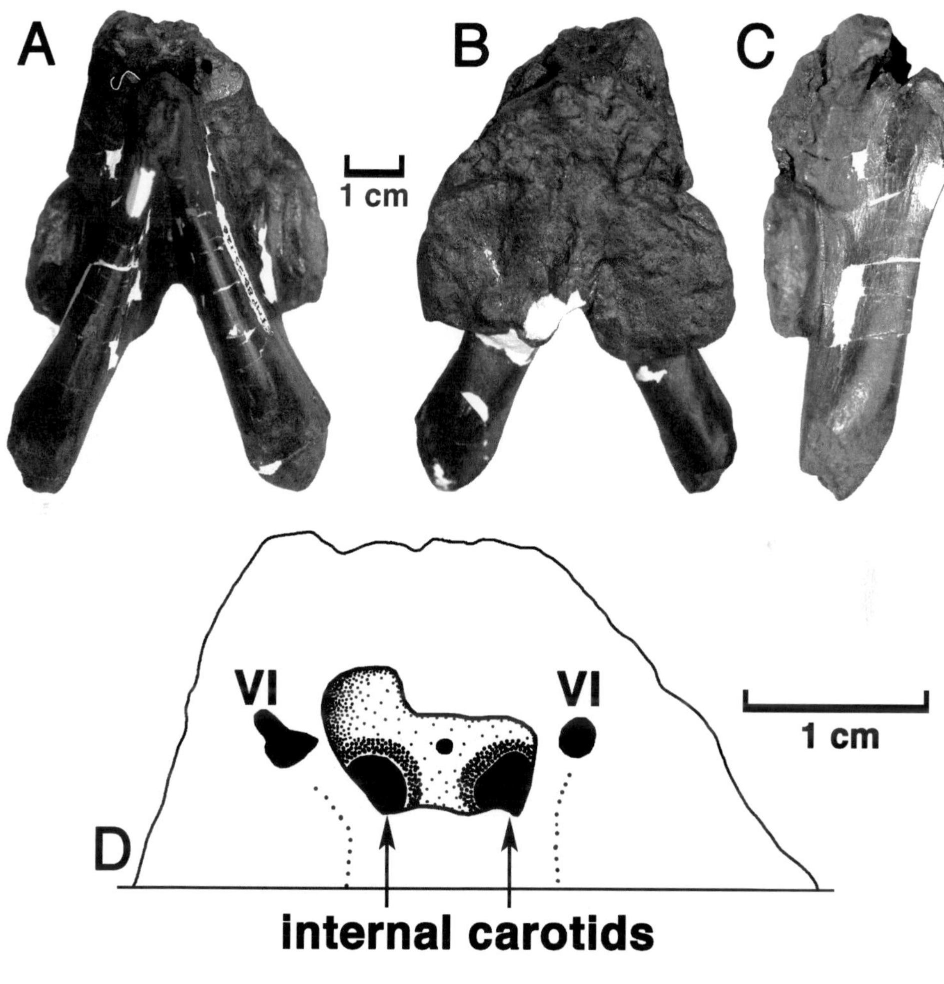

Cranial Nerves

Two braincases in the collection are particularly useful for understanding the cranial nerves. TMP 1989.55.517 (Fig. 39) is incomplete but uncrushed. The occipital condyle is 77 mm across and 71 mm high. The uncrushed foramen magnum is oval with a maximum width of 3.0 cm, and a height of 3.5 cm. It is split horizontally and has been prepared internally to produce an endocast similar to those described for *Anchiceratops* (Brown 1914*a*) and *Triceratops* (Forster 1990). The latex endocast of *P. lakustai* is 107 mm long from the cerebrum (where it is 43 mm wide) to cranial nerve XII. Computed tomographic scanning also led to the production of a digital endocast, which will be described in a separate paper in association with the vestibular apparatus (Witmer and Ridgely, this volume). The other braincase, TMP 1989.55.1243, had separated from the rest of the skull before burial, and even though it is distorted transversely (Fig. 40C) is informative in lateral view (Fig. 40B). As preserved, the condyle of this specimen is 5 cm wide, but in life, this dimension was probably about 6 cm, which is the height of the condyle. This specimen is smaller than the partial skulls with braincases in position (for example, the width of the occipital condyle of TMP 1989.55.188 is 71.5 mm and 78.5 mm high) and, probably, separated from the rest of its skull before burial.

The relatively large olfactory nerves (cranial nerve I) exit the front of the braincase through a single 14 mm wide opening in TMP 1989.55.517 (Fig. 39). The opening is partially subdivided into left and right tracts by a sagittal ridge.

The optic nerve (cranial nerve II) exits the braincase anterolaterally via a large, 5 mm diameter foramen that presumably passed through the orbitosphenoid bone, which is indistinguishably fused to the laterosphenoid. Cranial nerve III (oculomotor) passed through an opening that is almost as large as that of the optic foramen, and is positioned 7 mm posterodorsally above the pituitary fossa. Presumably it lies in the contact between the orbitosphenoid and laterosphenoid. Two small (circa 2 mm diameter) foramina pass through the laterosphenoid 15 and 23 mm dorsal (and slightly lateral) to the optic nerve foramen. One of these, probably the more posterior one, may have transmitted the trochlear nerve (cranial nerve IV), and the other probably transmitted a blood vessel.

Cranial nerve V (trigeminal) passed from the anterior end of the medulla through the foramen ovale. In *Triceratops* (Forster 1990), this foramen lies entirely within the prootic bone, but plesiomorphically, it passes between the laterosphenoid and prootic. The foramen is positioned slightly more than 1 cm dorsal, anterior, and lateral to a small foramen in the floor of the braincase that transmitted cranial nerve VI (abducens). Cranial nerve V divided almost as soon as it entered the bone (Fig. 39B, left side of neurocranium), the ophthalmic branch taking a more anterior course to emerge on the anterior wall of the laterosphenoid. A separate canal for the ophthalmic branch is also present in *Anchiceratops* (Brown 1914*a*). More laterally (Fig. 39B, right side of braincase), the maxillary and mandibular branches of the trigeminal nerve separated, one passing out of the braincase anteroventral to the other. Cranial nerve VI (abducens) opens on the anterior face of the dorsum sellae immediately lateral to the pituitary fossa.

The facial nerve (cranial nerve VII) entered a 4 mm wide foramen on the medial wall of the prootic region of the neurocranium, behind and above the trigeminal foramen. It emerged on the lateral wall of the braincase (Fig. 40B) in a slightly more posterior and ventral position than where it entered the braincase. The course of the acoustic nerve (cranial nerve VIII) cannot be seen, although the disparity in the sizes of the medial and lateral openings for the facial nerve suggests that at least one branch of cranial nerve VIII left the endocranium with cranial nerve VII.

The base of the vagus nerve (cranial nerve X) is marked by a 1 cm wide opening that passes laterally into the metotic fissure behind the fenestra ovalis. There are no indications of cranial nerves IX (glossopharyngial) and XI (accessory) leaving the braincase independently of cranial nerve X. Presumably, the glosspharyngeal emerged laterally through the metotic fissue while the accessory turned posteriorly with the vagus to exit the braincase through a large foramen lateral to the occipital condyle.

A line of three branches of the hypoglossal nerve (cranial nerve XII) sequentially leave the braincase posterior to the vagus exit. The smallest branch is the most anterior one, and the largest is the most posterior. Only one exit on each side is evident on the occipital surface, slightly dorsal to the foramen for cranial nerve X. Although the number of roots for the hypoglossal nerve can be variable within a species (Hopson 1979), both *Anchiceratops* and *Triceratops* (Forster 1996*a*) also have three per side.

Mandible

The lower jaw is represented by a large number of predentary and dentary bones representing a wide range of individual sizes. Also present are a few surangulars and articulars.

Predentary

Predentary bones range in length from 58 to 231 mm. The smallest example, TMP 1987.55.118 (Figs. 42D–42F) is 48 mm wide posteriorly and has a maximum depth of 25 mm. Although the large and small predentaries appear quite different, they have about the same length to depth ratio (0.35), and the apparent differences in shape are due to distortion of the larger specimens. The predentary displays a pronounced diagonal sulcus extending anteroventrally from near midlength (Fig. 42A). The sculpture of roughly parallel axial grooves and ridges is much more pronounced anterior to this sulcus than posterior to it. A second shallower sulcus, roughly paralleling the first and posterior to it, extends from the salient at the posterolateral edge of the predentary toward the ventral midline or joins with the first sulcus distally. The surface of the predentary is only lightly marked between the two sulci and is largely unsculptured behind the second. In artistic renditions of ceratopsians in the flesh, approximately all of the predentary is often represented as encased in a horny sheath. However, in *P. canadensis*, it appears that a sheath invested only the part of the predentary anterior to the foremost of the diagonal sulci.

Fig. 42. Pachyrhinosaurus lakustai, *predentaries. TMP 1989.55.1223 in ventral (A), occlusal (B), and left lateral (C) views and TMP 1987.55.118 in ventral (D), occlusal (E), and lateral (F) views.*

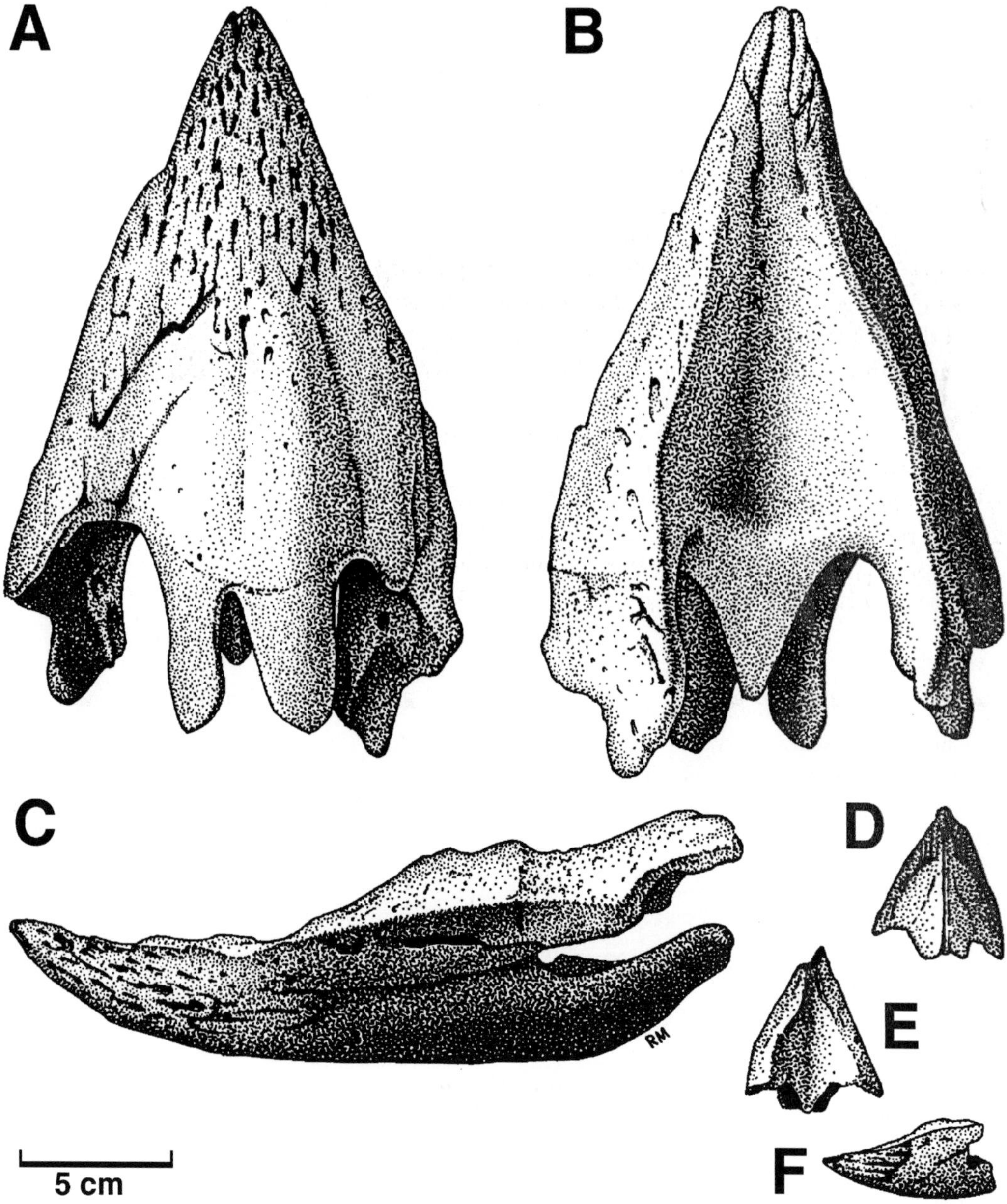

The dorsolateral surfaces of the predentary are broadly excavated longitudinally as in other ceratopsids. It is generally believed that this groove reflects a corresponding feature in the horny sheath, and it might be assumed that this received an occluding edge of the rostral. However, a corresponding ridge does not occur on the rostral bone, which is sharp edged and steeply vaulted. Except at their anterior tips, it is impossible to correlate any opposing features on the rostral and predentary. Thus, as in turtles, the upper beak probably contained more horny tissue than the lower, but it is impossible to be sure about the appearance and occlusion of the upper and lower beaks in the absence of these soft tissue structures.

Fig. 43. Pachyrhinosaurus lakustai *dentaries. (A) Left lateral view of a juvenile example (TMP 1987.55.253); (B and C) left lateral and medial views of adult individual (TMP 1987.55.129); (D, E, and F) ontogenetic series of TMP 1987.55.253, TMP 1986.55.97 (reversed) and TMP 1987.55.129.*

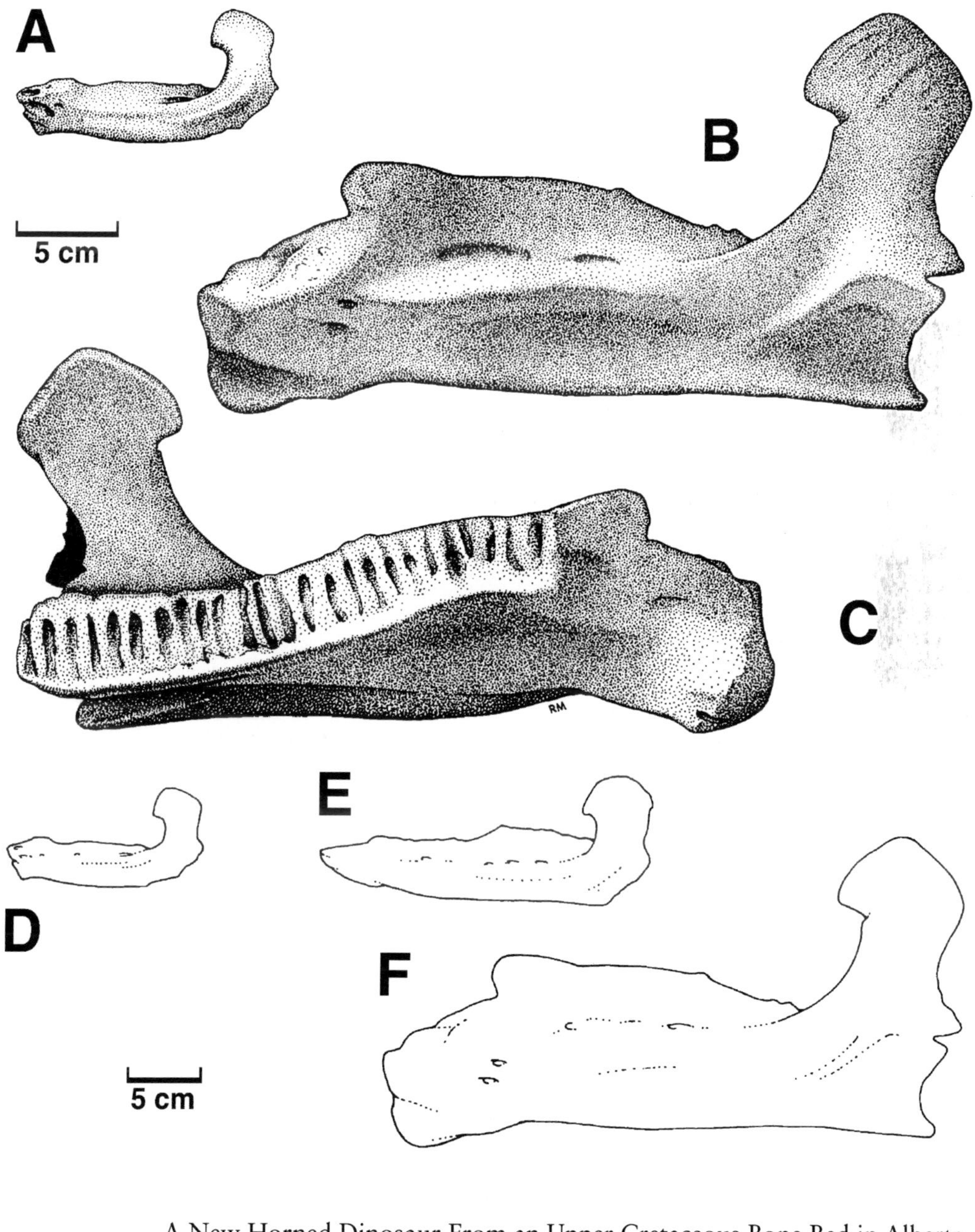

Dentary

More than 25 dentaries were collected in the Pipestone Creek bone bed, almost 60% of which are from the left side. A large (401 mm long) left dentary (TMP 1987.55.129, Figs. 43B, 43C) is almost complete; however, like all other examples in the collection, the thin lingual lamina that covered the dental magazine medially has been stripped away, and all teeth have been lost. The dentary is robust, deep, and possesses a relatively high coronoid process. It is 199 mm high at the coronoid process and the depth at midlength is a little more than 110 mm. The diastema between the tooth row and the posterodorsal process of the predentary is 66 mm long, and the distance from the top to the bottom edge of the bone in this area is 113 mm. This is some 15% wider than the unusually narrow (for a ceratopsid) space in the *P. canadensis* jaw from Scabby Butte (CMN 10645), suggesting that that specimen may be abnormal. The dental magazine was 290 mm long. Twenty-five or 26 alveoli are exposed, and this appears to be the total number and is comparable with other centrosaurines of this size. Sternberg (1950) estimated that there were 35 alveoli in the paratype of *P. canadensis*, a surprisingly large number for a short-faced ceratopsid, although some chasmosaurines (e.g., *Triceratops*) may have more than 40 tooth positions. An immature dentary 235 mm long (TMP 1986.55.97, Fig. 43E) contains about 20 alveolar grooves. A small example with a length of 140 mm (TMP 1987.55.253, Figs. 43A, 43D), probably had fewer grooves; however, because of the relative softness and spongy nature of juvenile ornithischian bone, the details of tooth groove morphology were obliterated by preburial weathering. Clearly, as in the maxilla, the number of tooth positions increased ontogenetically in *P. lakustai* as in other ceratopsians.

Angular

Several angulars have been recovered from both the right (TMP 1987.55.179, TMP 1989.55.528, TMP 1989.55.982, Fig. 44) and left (TMP 1989.55.892) sides. All are more than 11 cm long, which suggests that they were from adult individuals. They differ from other centrosaurines in that the lateral wing is dorsoventrally lower than it is anteroposteriorly long and has a sharp inflection separating it from the ventral surface. In all of these specimens, the lateral surface behind the dentary suture is pierced by two foramina, which vary in relative positions. The ventrally overlapping splenial suture is widest ventromedial to the overlapping dentary suture, tapers at midlength to a narrow contact, but expands again somewhat posteriorly (Fig. 44A).

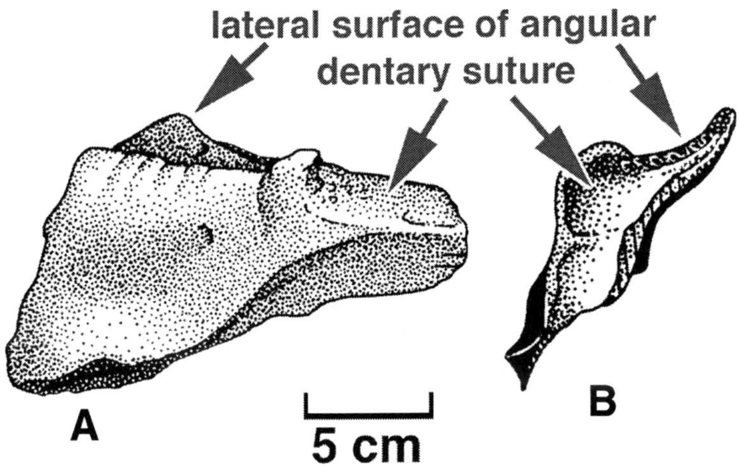

Fig. 44. Pachyrhinosaurus lakustai *right angular (TMP 1989.55.982) in ventral (A) and anterior (B) views.*

Surangular and articular

Co-ossified articular–surangular pairs are known from both the right (TMP 1987.55.32, 1989.55.969) and left (TMP 1988.55.223, Fig. 45; TMP 1987.55.293, TMP 1989.55.1285) sides. They have all suffered some dorsoventral crushing, but the surangular is comparable with that of a Scabby Butte specimen (CMN 10629) described by Langston (1975). It resembles the surangular of *Centrosaurus* (TMP 1986.18.39) more than CMN 10629 because the posteromedial edge of the ascending ramus is sharp and not emarginated as described in *P. canadensis*. The external, sculptured surface of the surangular is pierced by three or four vascular foramina behind the contact sutural

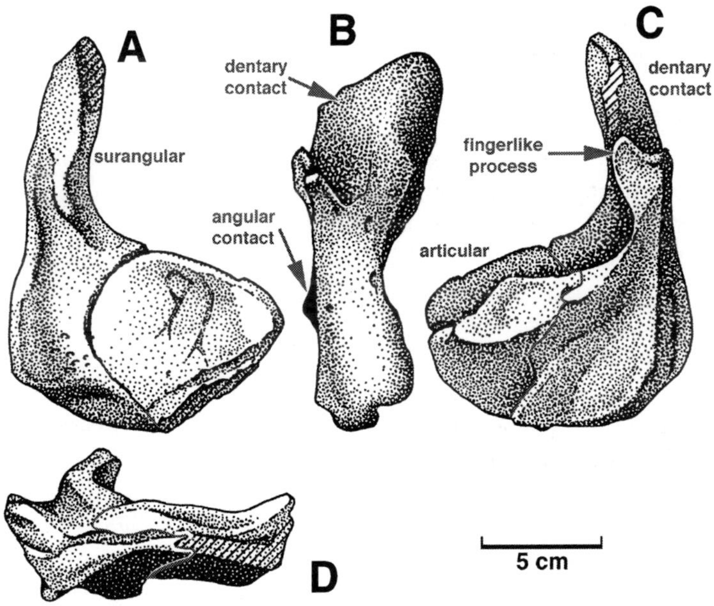

Fig. 45. Pachyrhinosaurus lakustai *left articular and surangular (TMP 1988.55.223) in posterodorsal (A), lateral (B), anteroventral (C), and posteroventral (D) views.*

C

surfaces for the dentary and angular as in *Centrosaurus*. The external surface has developed into a ridge and finger-like process between these two contact areas (Figs. 45B, 45C) and appears to be more strongly developed than in other centrosaurine specimens.

In addition to the articular–surangular pairs, there are three disarticulated articulars from smaller animals (TMP 1986.55.33 (left), TMP 2004.72.07 (left), TMP 2004.72.08 (right)). The smallest specimen is the most complete one and is only 37.4 mm long, including the weakly developed retroarticular process. The glenoid part of the articular forms the anterior half of the upper surface and is separated from the more posterior retroarticular process by a pronounced groove (in this and all of the more mature specimens). The articular surface is broad and convex dorsally in its medial region and tapers laterally where it becomes concave dorsally. The retroarticular processes of these specimens are less robust than those in *Centrosaurus* articulars. However, this may be related to the younger ages of the specimens at the time of death and to their distortion or loss in the more mature specimens.

Dentition

Teeth of ceratopsid dinosaurs have been well described by Hatcher et al. (1907), Tyson (1977), Sampson (1993), and others. Brown (1914*b*) described the teeth of *Centrosaurus* as having relatively higher central carinae than in *Triceratops*, which therefore resulted in relatively greater outer curvature. Langston (1967) described and illustrated a tooth from the Drumheller specimen. Numerous teeth were found in the Pipestone Creek bone bed (Fig. 46), although most were poorly preserved. With the exception of a single tooth in one maxilla (TMP 1989.55.855), all had fallen out of either maxillae or dentaries. Each maxillary crown has a strong, vertical ridge on the posterior half of the enameled labial surface. The distal end of a weaker, more posterior longitudinal ridge does not reach the margin of the crown in any of the unworn teeth. There are small but well-defined crenulations in the enamel along the margins of the labial surface. There is also a median, longitudinal ridge on the lingual surface of the maxillary crown. Depressions in the anterior and posterior surfaces of the base of the crown mark the contacts with adjacent tooth crowns. The posterior depression is the deeper of the two. Dentary teeth are similar to maxillary teeth except that the more heavily enameled surface is on the lingual surface, and the median ridge is more anterior than posterior. The horizontal ridge (cingulum) at the base of the enameled side of the crown is less pronounced in dentary teeth. The roots of mature teeth bifurcate in the typical ceratopsid manner. Teeth at the front of maxillary and dentary series have lower ridges and are single rooted (Hatcher et al. 1907). All of these characters are known in other ceratopsids and do not distinguish *P. lakustai* or other pachyrhinosaurs. Sternberg (1950) felt that the teeth of *P. canadensis* were relatively short and stout.

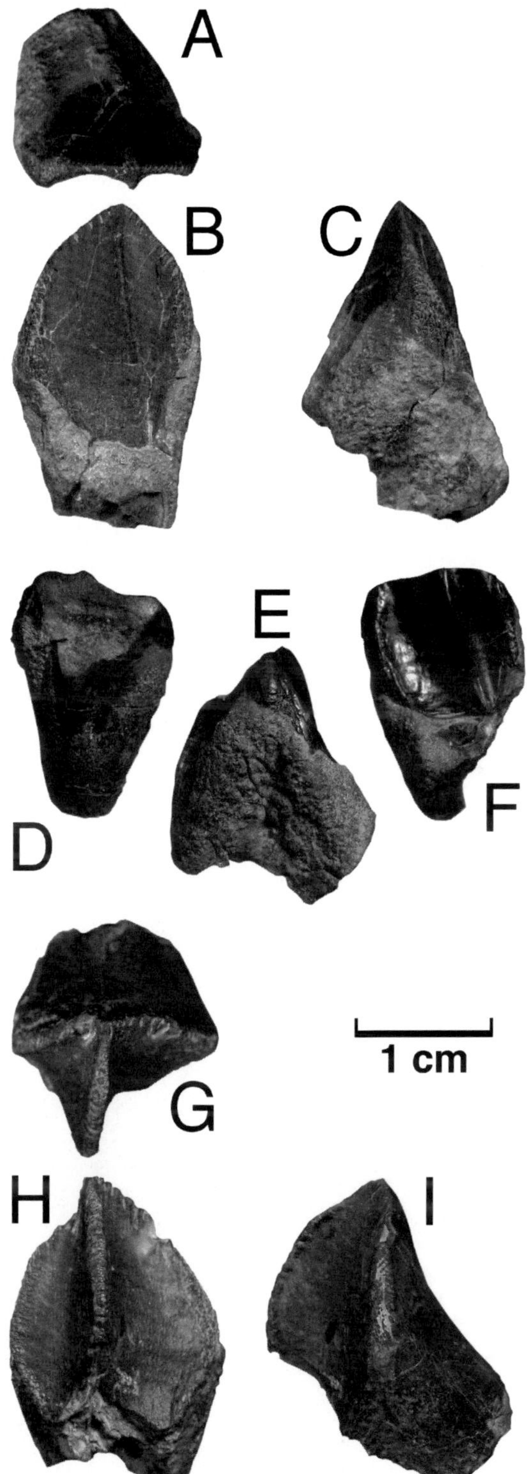

Fig. 46. Pachyrhinosaurus lakustai *teeth. Right maxillary tooth (TMP 1987.55.145) in distal (A), labial (B), and posterior (C) views; right maxillary tooth (TMP 1989.55.961) in lingual (D), posterior (E), and labial (F) aspects; and right dentary tooth (TMP 1989.55.833) in distal (G), lingual (H), and posterior (I) views.*

The numbers of maxillary and dentary teeth have been discussed in the context of those bones and are clearly ontogenetically variable as in other ceratopsians (Dodson et al. 2004). The number of teeth stacked within each alveolar groove is not known for *P. lakustai*, although there are as many as five in each stack in *P. canadensis* (Sternberg 1950). Almost certainly the number of stacked teeth within an alveolus was also ontogenetically variable.

The majority of teeth recovered have vertical or subvertical wear facets (Fig. 46D), the angle of which is determined by the position in the jaw and the degree to which the tooth had erupted at the time of death (Lull 1908; Tyson 1977).

Hyoids

Two hyoid elements (Fig. 47) were recovered from the Pipestone Creek bone bed. TMP 1989.55.691 is an almost complete, ossified section of a left first ceratobranchial (see Colbert 1945 for a discussion of hyoid terminology) that is 155 mm long. In all features, it is

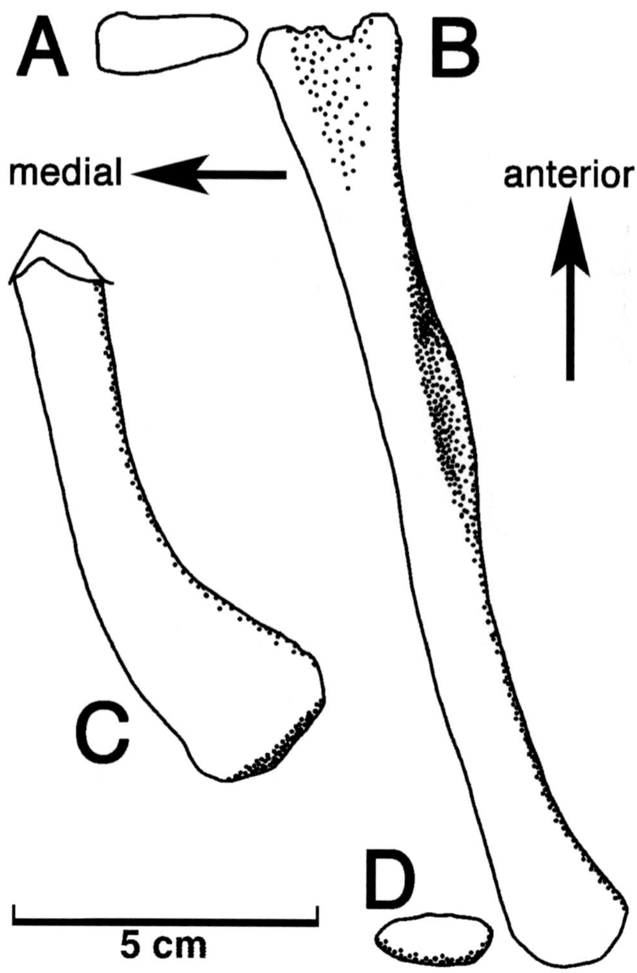

Fig. 47. Pachyrhinosaurus lakustai *hyoids. TMP 1989.55.691 in anterior (A), ventral (B) and posterior (C) views. (D) Posterior end of hyoid TMP 1989.55.1338 in ventral aspect.*

almost identical to a slightly larger (168 mm) ceratobranchial I of *Centrosaurus apertus* that was found in position between the mandibles of ROM 767 and described by Parks (1921). The somewhat flattened shaft is 12 mm wide (compared with 16 mm in ROM 767), with a posterior end 18.6 mm across, and a squared-off anterior end that is 24.2 mm wide. The posterior end of a larger first ceratobranchial (TMP 1989.55.1338, Fig. 47C) is broader in relation to shaft width and curves more strongly laterally. Hyoid elements have been described in other ceratopsians for *Leptoceratops* (ceratobranchial I; Sternberg 1951), *Protoceratops* (ceratobranchial II; Colbert 1945) and *Psittacosaurus* (first and second ceratobranchials; Colbert 1945). The unusual curved hyoid described by Lull (1933) for *Triceratops* presumably represents more extensive ossification of the first ceratobranchial.

Osteoderms

Epoccipital and epijugal elements presumably of osteodermal derivation have already been noted. In addition, there is a small isolated ossicle (Fig. 35) from the bone bed. This may be an epoccipital, an epijugal, an unattached nub related to the parietal horns, or even perhaps an ossicle from one of the crater-like cups randomly distributed on the narial bridge and the sides of the face.

Pathologies

A small number of interesting cranial palaeopathologies have been discovered in the Pipestone Creek bone bed material (Tanke 1989, 2005, 2006; Rothschild and Tanke 1992, 1997; Tanke and Rothschild 1997, 2002; Tanke and Farke 2003, 2007). Resorption pitting, which is sometimes considered to have pathological origins, is not considered here because it appears to have normal biological significance related to ontogeny (Sampson et al. 1997; Tanke and Farke 2006).

A partial skull of an adult pachyrhinosaur (TMP 1989.55.1234, Fig. 7C) has a number of pathologies, including a large penetrating lesion that affected the right side of the face plus ventral edge resorption and medial surface lesions of the left squamosal. The first lesion was originally interpreted as a horn thrust injury (Rothschild and Tanke 1996, 1997), but more recent interpretations (Tanke and Farke 2006) suggest more benign origins such as bone resorption or undetermined disease processes.

The rostral of a partial skull (TMP 1989.55.188, Fig. 8B) has a semicircular notch in the left ventral margin just behind the tip of this element. Elsewhere on the skull, there are broken and rounded bone edges incurred from taphonomic processes, but no similar damage occurs on the snout region suggesting the rostral notch truly was an injury, possibly resulting from biting a hard food item. The broken bone edges are now rounded, but it is unclear whether this represents healing, water wear, or polishing during sediment compaction. An identical condition affects the rostrum of a pachyrhinosaur from Dinosaur Provincial Park (TMP 2002.76.1).

Several parietal injuries, some of remarkable severity, are known. In adult centrosaurines, the posterior transverse ramus of the parietal is a greatly thickened and complex element of considerable strength. However, in younger animals, the transverse ramus was a thin, simple, and fragile structure (Dodson 1984, 1986; Dodson and Currie 1988; Sampson et al. 1997) that was more likely to be injured. Two subadult parietals exhibit pathological features. An unidentified active disease process resembling osteomyelitis eroded a large region of the dorsal and right lateral surfaces of the median parietal bar of TMP 1988.55.90. This removed the marginal undulations that, upon maturity, would have developed into the medial spike complex. The second specimen (TMP 1989.55.125) consists of the posterior parietal bar section with a healing transverse fracture on the parietal bar. Along this healing fracture, instead of continuing on a straight plane, the bone is flared ventrally and, to a lesser degree, dorsally with smooth rounded edges. The face of the break shows a smooth undulating surface, nearly identical in texture to a pseudoarthrosis surface on a neural spine of a midcaudal vertebra of a hadrosaur (TMP 1989.36.321). The injury was apparently noninfectious, although ventrally there is a 2 cm wide bone texture change posterior to the break.

One adult parietal shows a massive injury and unusual posttrauma healing. TMP 1987.55.210 (Fig. 32C) had lost the left frill spike before death, and extensive bone remodeling resulted in a strongly asymmetrical frill. The left side of the frill ends anteriorly in a smooth rounded point suggesting a pseudoarticulation. Ventrally, two round and roughly textured scars mark the horncore base and contact point for the left lateral ramus of the parietal. These two scars may represent pseudoarticulation surfaces, but this cannot be determined with certainty.

Another adult parietal (TMP 1989.55.1085; Fig. 33H) includes both P3 frill spikes, the left one with the usual outward and forward curvature, whereas the right one (on the misshapen side of the frill) points upward and forward. In section, the horncore is a flattened oval rather than round, although this may represent individual variation. Both TMP 1987.55.210 and TMP 1989.55.1085 have unusual shapes, but there are no other anomalies (such as extra growths, diseased bone, or disrupted bone surface textures), suggesting that the injuries occurred long before death and had time to heal. If the parietals were injured when the animals were young, the strong positive allometric growth of the frill would explain the asymmetry of these parietals. However, the presence of the base of a frill spike in TMP

1987.55.210 suggests that the parietal was injured when the individual was fully grown and presumably sexually mature. Adult parietal asymmetry is also known from a single *Centrosaurus* example (TMP 1964.5.191) from southern Alberta.

Langston (1975) described a large sinus on one side of a parietal frill of *P. canadensis* from Scabby Butte. No similar cavities are present in frills from Pipestone Creek in which the interiors of both the parietal and the spikes are filled with cancelous bone. No lumen exists. Therefore, the condition in the Scabby Butte frill appears to have been pathological or diagenetic in origin.

TMP 1986.55.111 (Fig. 8E) is the an adult skull that has a smooth, flattened lobe of bone (40 mm long and 37 mm wide) with a rounded distal end on the underside of the anteroventral edge of the left squamosal. Viewed externally, the lobe protrudes for a distance of approximately 20 mm below the ventral edge of the squamosal. The bone surface of this feature and its surrounding area is smooth, which suggests this may be a growth anomaly rather than an injury or indication of disease.

TMP 1987.55.101 is a quadrate from an adult specimen that has three anomalies on the distal end. The distal articulating surface is concave rather than convex, and there is a blunt, heavily constructed, smooth peg protruding ventrally from the articular surface of the medial condyle. Above this and slightly medially, a roughly circular (20 mm) hole penetrates up to 13 mm into the anterior surface. These three features are probably related, but their etiology is unknown. The peg-like growth must have disrupted the action of the lower jaw, because it would have directly contacted the articular, possibly causing eburnation.

A second quadrate from an adult (TMP 1989.55.1072) has two pathologies. There is a circular depressed area on the articulating surface of the medial condyle. The texture of the floor of the depression is identical to that of TMP 1988.55.90, an immature pachyrhinosaur parietal bar, and suggests an infectious process. The second pathology is a 10 mm deep crack, which appears to be a partially healed fracture, that extends diagonally between the articular surfaces of the two distal condyles. At midlength, a thin wall of bone separates the crack into two sections.

Table 6. Data matrix of character states used for the phylogenetic analysis.

Species	Morphological character																				
	1	2	3	4	5	6	7	8	9	10	11	12	13	14	15	16	17	18	19	20	21
Protoceratops andrewsi	0	0	0	0	0	0	0	0	0	0	0	0	0	0	1	N	0	N	N	N	N
Zuniceratops christopheri	?	0	0	?	?	0	1	?	1	?	0	?	?	?	0	0	1	0	0	1	1
Albertaceratops nesmoi	?	?	?	?	?	1	?	?	1	1	2	1	1	?	0	0	1	0	0	0	1
Centrosaurus apertus	0	1	1	0	0	1	1	1	1	0	1	1	1	1	0	1	1	1	1	0	1
Styracosaurus albertensis	0	1	1	0	0	1	1	?	1	0	1	1	1	1	0	1	1	1	1	0	N
Achelousaurus horneri	0	1	1	0	0	1	1	1	1	1	3	1	1	0	1	2	2	N	1	0	N
Einiosaurus procurvicornis	0	1	1	0	0	1	1	1	1	0	1	1	1	0	1	2	1	2	1	0	N
Pachyrhinosaurus canadensis	0	1	1	0	0	1	1	1	2	1	3	1	1	?	1	2	2	N	1	0	N
Pachyrhinosaurus lakustai	0	1	1	0	0	1	1	1	2	1	3	1	1	0	1	2	2	N	1	0	N
Chasmosaurus belli	1	2	2	1	1	0	0	0	1	0	1	0	1	0	0	?	1	0	0	1	1
Triceratops horridus	1	2	2	1	1	0	1	0	1	0	1	0	0	0	0	0	1	0	0	0	0

Note: Morphological characters and character states are defined in Appendix A.
Character states are as follows: 0, primitive state; 1, 2, 3, derived character states; ?, missing data; N, not present.

Discussion

Phylogenetic analysis

Sampson (1995) conducted a phylogenetic analysis of centrosaurine relationships that included material from the Pipestone Creek Bone bed, basing his information both on an earlier version of this paper and examination of the specimens at the Royal Tyrrell Museum of Palaeontology. A subsequent examination of generic-level relationships of ceratopsids by Dodson et al. (2004) paired *Pachyrhinosaurus* with *Einiosaurus* based on two synapomorphic characters. An analysis by Ryan and Russell (2005) hypothesizes a clade of (*Einiosaurus* + (*Achelousaurus* + *Pachyrhinosaurus*)) supported by one unambiguous (nasal horn core longer than high) and one ambiguous (P2 process) character.

For the purposes of the phylogenetic analysis of the Pipestone Creek ceratopsid, a species-level analysis was conducted of the Centrosaurinae (seven ingroup taxa), a subfamily long established as a monophyletic taxon (Lambe 1915; Sternberg 1949; Lehman 1990;

22	23	24	25	26	27	28	29	30	31	32	33	34	35	36	37	38	39	40	41	42	43	44	45	46
0	0	0	0	0	0	0	0	0	0	0	0	0	?	0	0	0	0	0	0	0	0	0	0	0
?	?	?	?	?	?	?	?	?	?	?	?	?	?	?	?	?	1	0	0	0	0	?	?	1
?	1	1	?	1	1	1	2	0	0	0	3	3	?	1	1	1	2	1	1	1	1	?	?	?
1	1	1	1	1	1	1	2	3	3	3	1	0	1	1	1	1	2	1	1	1	1	0	0	1
1	1	1	1	1	1	1	2	2	3	2	2	2	1	2	1	1	2	1	1	1	1	0	0	1
1	1	1	1	1	1	1	2	0	0	2	2	2	1	1	1	1	2	1	1	1	1	0	0	1
1	1	1	?	1	1	1	2	0	0	1	2	1	1	1	1	1	2	1	1	1	1	0	0	1
1	1	1	1	?	1	1	2	0	0	3	2	3	1	1	1	1	2	1	1	1	1	0	0	1
1	1	1	1	2	1	1	2	0	0	3	3	3	?	1	1	1	2	1	1	?	1	0	0	1
0	0	0	0	0	1	2	1	1	2	1	1	0	0	0	0	0	2	1	1	1	1	1	1	2
1	0	0	0	0	0	2	1	1	1	1	1	0	0	0	0	0	2	1	1	1	1	1	1	2

Dodson and Currie 1990; Sampson 1995; Penkalski and Dodson 1999; Dodson et al. 2004; Ryan and Russell 2005; Ryan 2007). All currently accepted centrosaurine species were included other than *Centrosaurus brinkmani* and *Centrosaurus nasicornis* (neither of which is as well known as *Centrosaurus apertus*). *Zuniceratops christopheri* and *Protoceratops andrewsi* were selected as successively more basal ceratopsian out-groups. *Chasmosaurus belli* and *Triceratops horridus* were included as well understood members of the Chasmosaurinae, the phylogenetically most important out-group. The data matrix (Table 6) was derived primarily from Sampson (1995), Dodson et al. (2004), and Ryan (2007) but was extensively revised to incorporate new information gleaned from this and other studies. Eleven taxa were included in the analysis, and 46 parsimony informative characters were utilized. It was constructed using MacClade 4.08 (Maddison and Maddison 2005), which was also used to produce Fig. 48. The analysis was run using the beta version of PAUP 4.02b10 (Swofford 2002). The branch and bound search method produced a single tree (tree length = 90, consistency index = 0.8333, homoplasy index = 0.1667, retention index = 0.8193, rescaled consistency index = 0.6827).

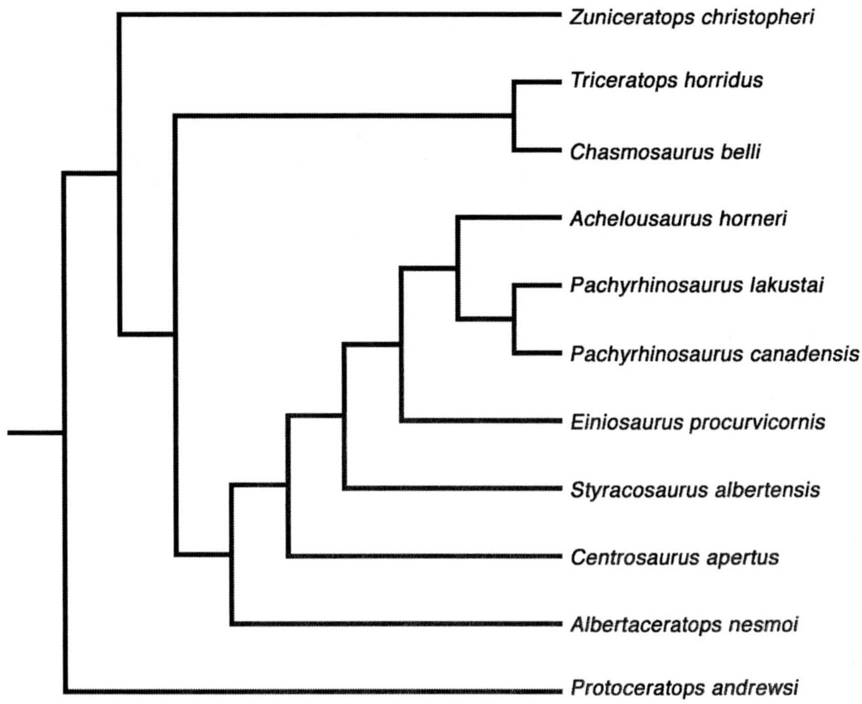

Fig. 48. Cladogram depicting the strict consensus tree obtained from the phylogenetic analysis of 11 ceratopsian taxa with an emphasis on Centrosaurinae.

Pachyrhinosaurus lakustai and *P. canadensis* are closely related to each other, form a crown group within the Centrosaurinae, and would be considered as congeneric in a traditional analysis. The constraint of monophyly required *Achelousaurus* and *Einiosaurus* to be treated as genera separate from *Pachyrhinosaurus* (Sampson 1995) to prevent the creation of a paraphyletic taxon. It is possible that the descriptions of new, closely related pachyrhinosaur taxa from Alaska (Fiorillo 2004) and Alberta (Ryan et al. 2006) may force the establishment of a new genus for the pachyrhinosaur from the Pipestone Creek Bbone bed. However, it is also possible that one or both of the undescribed animals will turn out to fall within the range of variation of the existing pachyrhinosaur species.

Pachyrhinosaurus lakustai is clearly closely related to *P. canadensis* but differs in being approximately 15% smaller at maturity, in having a relatively smaller rostral, in having a shorter, more robust premaxillary process that projects straight forward into the nares, in having a relatively shorter nasal boss that does not contact the supraorbital bosses (the bosses are separated by a smooth T-shaped groove), in having a parietal bar that curves slightly upwards rather than having an essentially flat profile in lateral view, in having one to three median horns near the base of the midparietal bar in subadult and adult stages, and in having a lateral parietal spike (P3) that twists anteroventrally instead of dorsolaterally.

Pachyrhinosaurus lakustai is easily distinguished from the geological older *Achelousaurus horneri* by its relatively larger nasal boss and by the anteroventral twist of its P3 spike.

Ontogeny of the cranial bosses

The presence of cranial bosses and other ornamentations only in larger examples of *Pachyrhinosaurus* and the more conventional centrosaurine cranial morphology exhibited by smaller individuals bespeaks some exceptional ontogenetic changes in this taxon. That there is little difference in the size of skulls with and without bosses indicates that such changes occurred rapidly and constituted a sort of metamorphosis whose onset may signal the achievement of sexual maturity. The expression of all cranial ornamentations was possibly more extreme in one sex than the other. We offer the following inferences about the sequence of events involved in the metamorphosis that, although shared with *Achelousaurus* and other centrosaurines to some extent, appears otherwise unique amongst dinosaurs.

The hornless nasal boss of *P. lakustai* is almost certainly derived ontogenetically from the small demihorned nasals. Growth of the nasals from the smallest example to approximately two-thirds of adult size is essentially isometric (Fig. 15). However, at some point beyond the two-thirds adult stage, the nasals increased rapidly in width and thickness. The demihorns, instead of coalescing to produce a horn core like that of *Centrosaurus*, remained divided and migrated apart with a massive addition of bone on the opposing (medial) surfaces of the nasal bones. As the breadth of the skull increased, remnants of the demihorns remained topographically at the edge of the cranial roof. The transverse expansion of the roof is foreshadowed in the deflected triangular area on the posterior side of the demihorn, which is seen to assume a more transverse position from the smallest to the two-thirds stage nasals. The lateral edge of the sulcus bounding the deflected triangular area grew upward rapidly to form the rim of the boss, and each demihorn was ultimately largely engulfed by thickening of the nasal until, as a relict, it is recognizable only as a blunt horn-like elevation along the rim. With growth, the so-called flattened bulge on the nasal near the base of the demihorn may have joined with its mate to produce the spout-like pommel of the adult skulls.

How long this transformation may have taken is unknown, but once adulthood was attained continuing growth augmented the basic structure. Thus, the increased thickness of the skull roof, elevation of the boss, and reduction of its rim in *P. canadensis* skulls from Scabby Butte, and the virtual absence of the elevated rim in the Drumheller skull (if not a postmortem effect), may indicate that these individuals were in a later stage of life than the *P. lakustai* examples at Pipestone Creek.

The suggestion that the boss developed from an erect horn core that subsequently extended forward over the dorsal premaxillary region as postulated by Sampson (1995) in *Achelousaurus* is not supported by available *P. canadensis* material. There is no evidence that a nasal horn core became procurved like that of *Einiosaurus* at any ontogenetic stage in *Pachyrhinosaurus*. Hence, a postulated fusion with the underlying nasal bone in *Achelousaurus* is not valid for *Pachyrhinosaurus*.

Similar rapid changes at about the same time in life probably account for the transformation of the postorbital bones into supraorbital bosses. The ridge-and-groove field may represent a region of rapid growth that was mobilized at a late juvenile stage. Ridges and grooves appear as parallel diagonal wrinkles on the medial side of the juvenile postorbital horn. The ridges developed first with the spaces between them being filled in later, producing a multilayered plicated pattern whose axes of breadth lay normal to the upwardly facing top of the boss. Support for this idea is found in the geologically older *Achelousaurus horneri* in which the supraorbital bosses, "possess several deeply excavated ridges, which are oriented approximately posteromedially, are several millimeters thick at the base, and thin dorsally" (Sampson 1995). In *A. horneri*, the ridge-and-groove field is widely exposed laterally as well as dorsally, cortical bone not having developed to the extent seen in *Pachyrhinosaurus*.

Sexual dimorphism

Although the sample size of prepared skulls is relatively small, there are a number of features in mature skulls that may represent sexual dimorphism, which has been already proposed in *Protoceratops* (Brown and Schlaikjer 1940; Kurzanov 1972; Dodson 1976), *Centrosaurus* (Dodson 1990), *Chasmosaurus* (Lehman 1989, 1990), *Einiosaurus* (Sampson 1995), *Styracosaurus* (Dodson 1990), and *Triceratops* (Ostrom and Wellnhofer 1986).

Early in the study of the pachyrhinosaur material from Pipestone Creek, two morphologies of nasal boss had been identified (Sampson et al. 1997)—the dorsal surface was convex in one form and supposedly concave in the other. Subsequent preparation and analysis of the nasal bosses revealed that the deeply concave morph invariably lacks the textured surface bone. It seems likely that, once the exterior surface was damaged postmortem, then the spongy interior bone would have disintegrated to leave a crater in the dorsal surface of the nasal boss. Another possibility is that highly spongy bone of the boss was a source of calcium for females when they were laying eggs, just as some theropods were able to utilize medullary bone (Schweitzer et al. 2005). At the time that the female was laying eggs, the structure of the spongy internal bone of the boss may have been weakened enough for it to collapse postmortem, leaving the deeply concave boss. It is necessary to do histological examinations of each of the mature nasal bosses to test these and other hypotheses. However, the absence of deeply concave bosses in other pachyrhinosaur species would also suggest that it is more parsimonious to assume that those from Pipestone Creek are preservational artifacts.

Another possible sexually dimorphic character can be seen in the array of spikes on the frill, especially those on the midline bar of the parietal. Although the evidence suggests that roughly one-half of the mature specimens each have one or more midline spikes, the sample size is too small to be sure that this feature is sexually dimorphic rather than individual variation in the ontogenetic timing of spike acquisition. As cautioned by Sampson et al. (1997), without large, statistically significant samples, it is difficult to differentiate amongst ontogenetic, individual, and sexual variation. Collection and preparation of additional parietal midline bars and (or) complete skulls, plus histological studies to potentially determine ontogenetic ages, might ultimately show that there is sexual dimorphism in the arrangements of parietal spikes. However, the evidence presently at hand can only be considered as an indication that ceratopsians may have been sexually dimorphic in their horns and spikes, just as many artiodactyls are today.

Function of nasal and supraorbital bosses

All ceratopsids have horn cores that were probably covered by keratinous sheaths. The absence of tapering horn cores in *P. canadensis* led Sternberg (1950) to speculate that this animal had a low mound of keratin capping the nasal boss to form a battering ram, and he believed that it did not appear likely that there was a pointed horn. However, Sternberg (1950) also noted the heavily textured surface of the nasal boss, and the presence of a lateral and posterior sulcus at the base of the nasal boss, which suggested to him the presence of a huge horny mass. Langston (1967) discussed the possibility that the boss supported some sort of bony nasal horn that had become detached or broken, but eliminated the idea because the lack of fusion of such a horn in all known specimens seemed unlikely, as did the fact that isolated ossified horns had never been found amongst the abundant *Pachyrhinosaurus* material. However, the similarity of the pachyrhinosaur nasal boss to the nasal bosses of rhinoceroses (especially *Elasmotherium*) suggests that it may have supported a keratinous horn that would not have been fossilized (Currie 1989). Sampson et al. (1997) questioned this hypothesis by asking why ceratopsians would evolve two different types of nasal horn. However, a nonosseous epidermally derived horn would have been lighter and tougher than a horn composed of a bone core with a keratinous sheath and would have been less susceptible to serious injury. Hieronymous et al. (2006) used osteological correlates to test various hypotheses of epidermal covering of the bosses of *Pachyrhinosaurus* and concluded that the most parsimonious hypothesis is that the boss was covered with a papillary horn that was probably almost flat dorsally. Further analysis of the fossil material at hand is unlikely to solve the mystery of the shape and structure of the nasal boss covering, let alone its function.

The ontogenetic development of the supraorbital boss mimics that of the nasal boss in *Pachyrhinosaurus*, whereas phylogenetically the development of supraorbital bosses precedes the evolutionary appearance of the nasal boss in ceratopsids. Mature specimens of *Centrosaurus* and *Styracosaurus* often show evidence of resorption of supraorbital horns but never have any indication of nasal horn reduction. *Einiosaurus* has a strongly procurved nasal horn at maturity and low, dorsally concave supraorbital horns. *Achelousaurus* has both nasal and supraorbital bosses, the latter with distinct lateromedial ridges and furrows as in *P. lakustai* (Figs. 20, 24). The nasal and supraorbital bosses are well separated from each other in *Achelousaurus horneri* and *P. lakustai*. Whereas the nasal boss is relatively short (Table 4) and is only marginally longer than the supraorbital boss in the former, it is significantly longer in the latter. In *P. canadensis*, the nasal boss is also significantly longer than the supraorbital boss, but the separation between them is reduced or lost. This is also true for the pachyrhinosaur specimens from Alaska. Although the boss clearly increases in relative size ontogenetically in *Pachyrhinosaur lakustai*, there is no indication amongst the adult specimens to suggest that the closure of the gap between the nasal and supraorbital bosses is controlled by the size of the specimens. The trend within this lineage of centrosaurine ceratopsids is a move away from the three facial horns with bone cores to some kind of keratinous structure with three centers (nasal plus two supraorbital), to a larger structure with a single base that extends from above the naris to above the orbits.

The function of ceratopsian horns and frills have received considerable attention in recent years, and our purpose here is not to review and elaborate on a topic that has been well covered elsewhere (Spassov 1979; Dodson 1996; Sampson et al. 1997; Dodson et al. 2004). However, the *P. lakustai* material does permit insight into certain aspects of the various discussions of nasal boss function. Whereas many authors (including Farlow and Dodson 1975; Farke 2004) have discussed the possibility that the ceratopsian horns and bosses functioned for defense and intraspecific combat, most recent papers accept that the main function of these structures was for intraspecific identification for purposes of reproduction (Farlow and Dodson 1975; Dodson and Currie 1990; Sampson et al. 1997; Dodson et al. 2004). The nasal boss of *P. lakustai* is elongate; if it were used as a "battering ram" in interspecific or intraspecific combat, then the dorsal surface would have been the one that contacted the potential foe. This would have been problematic in that any hard contact by the nasal boss would have rotated the head downwards around an arc centered on the occipital condyle, and this in turn would have then brought the spikes on the parietal bar into contact with whatever the nasal boss had contacted. If this happened to be another individual of the same species, the parietal bar spikes would have potentially inflicted a serious puncture in the head of the opponent. This would not have been the case if the epidermal structure on top of the nasal boss were higher than the parietal spikes, because the contact surface would not necessarily be the top of the boss in this case. If anything though, the presence of the midline, parietal spikes argues in favor of the nasal boss (and all of the parietal–squamosal ornamentations and epoccipitals) being display structures rather than offensive or defensive weapons.

Origin of the *Pachyrhinosaurus* bone bed on Pipestone Creek

Monodominant ceratopsian bone beds are relatively common in southern Alberta (Langston 1975; Currie and Dodson 1984; Eberth and Getty 2005) and have been found in Alaska (Fiorillo and Gangloff 2003), Montana (Rogers 1990; Sampson 1995), and Texas (Lehman 1989, 1990). *Pachyrhinosaurus* is the dominant genus of bone beds at Scabby Butte (Langston 1975), along the Colville River in Alaska (Fiorillo 2004), and at Pipestone Creek. It is now generally accepted that these massive accumulations of hundreds or even thousands of individuals represent catastrophic kill-offs of gregarious groups or herds of ceratopsians. Mechanisms proposed for the genesis of the monodominant bone beds include river flooding (Currie 1981; Currie and Dodson 1984), coastal inundation (Eberth and Getty 2005), and drought (Rogers 1990). The bone bed at Pipestone Creek is contained within fine-grained sediments in overbank deposits, and the bones show relatively poor sorting and hydraulic reworking. The bones were deposited in a single but dense layer less than 1 m in thickness. Virtually all bones of the skeleton are disarticulated; because of the large number of individuals involved, it is impossible to determine whether or not there are associations of single skeletons within the bone bed. There is little evidence of tooth marks, trampling, or weathering, although associated shed theropod teeth suggest that there was some preburial scavenging. Although the taphonomic evidence suggests that the burial occurred over a relatively short period and that all the individuals were from the same gene pool, a possible cause of death is not evident. Further excavation at Pipestone Creek may eventually produce some clue to the origin of the bone bed.

The presence of large herds of *Pachyrhinosaurus* in Alberta and Alaska has suggested the possibility to some (Currie 1989, 1992; Dodson 1996) that this dinosaur migrated on an annual cycle between these two regions. Although arguments in favor of the concept of migration (Hotton 1980; Spotila et al. 1991) are compelling for large dinosaurs, considerable research needs to be done to show that this in fact happened. Dodson (1996) pointed out that, before this hypothesis can be addressed in any manner, it is necessary to identify the Alaska species. Superficial examination of a specimen from the North Slope (UCMP field number 88H8-4-1) shows that *P. lakustai* is not the animal that has been discovered in Alaska, but it could not be determined whether or not the specimen represents either *P. canadensis* or a new taxon.

Summary

Pachyrhinosaurus lakustai sp. nov. is represented by the remains of more than 27 individuals recovered from a densely packed bone bed on Pipestone Creek near Grande Prairie, Alberta. It differs from *P. canadensis* in numerous ways, some of which are transitional states between the conditions seen in *Achelousaurus horneri* and *P. canadensis*. All three pachyrhinosaur species have nasal and supraorbital bosses rather than the horns that are present in less derived but related forms like *Centrosaurus apertus*. The nasal boss is relatively small in *Achelousaurus horneri*, significantly larger in *P. lakustai*, and large enough to contact the supraorbital bosses in *P. canadensis*. This transition seems to match the temporal sequence of the three species. However, in other ways, *P. lakustai* seems to be more derived than *P. canadensis*. Autapomorphies include an unusual series of bumps on the rostral and premaxillary bones that is referred to here as a "rostral comb" and a large anterolaterally curving epoccipital horn on the posterolateral edge of the parietal section of the frill. *Pachyrhinosaurus lakustai* also seems to have been about 20% smaller at maturity than *P. canadensis*.

Pachyrhinosaurus lakustai remains cover a wide range of ontogenetic development, and the largest individuals are almost 3 times the linear size of the smallest ones. Cranial morphology undergoes one of the most remarkable ontogenetic transitions documented for any dinosaur: the conical nasal and supraorbital horns of the juveniles transform into the broad and massive bosses of the adults, and large spikes develop on the middle parietal bar and on the back of the frill. Coupled with the rapid development of these cranial structures is an amazing degree of bone absorption that creates craters where convex structures had been and huge sinuses within the supraorbital bones. Bivariate comparisons of cranial structures in *Pachyrhinosaurus* provides detailed information on the allometric growth of this (and related animals). For example, it is clear that, in comparison with foramen magnum width, there is strong positive allometry in the growth of the occipital condyle. Of courset this is not unexpected because the width of the spinal chord is correlated with the negative allometry of growth of brain size, whereas the condylar size itself is correlated with the strong positive allometry of the overall weight of the skull that it needs to bear.

It is not possible at this time to determine what the caused the death of the many individuals of *P. lakustai* that have been found in the monodominant Pipestone Creek bone bed. The minimum number of 27 individuals is sure to increase whenever excavation resumes at the quarry. Plans are underway to construct a new facility at the site to display a section of the bone bed in situ, which extends for at least another 70 m from the existing quarry (Fanti and Currie 2007).

The temporal and geographic distribution of rich bone beds of *Pachyrhinosaurus* and the other centrosaurine ceratopsians has enormous potential for investigations of rates and modes of evolutionary change; of individual, ontogenetic, or sexual variation; and even of behaviour. Whereas such studies were unheard of only three decades ago, the discovery of bone beds like the one at Pipestone Creek have dramatically changed our perceptions of ceratopsians and other dinosaurs. The boreal distribution of herds of *Pachyrhinosaurus*, an animal with florid elaborations of cranial morphology suggestive of some modern animals living in rich tropical ecosystems, reinforces evidence that dinosaur diversity was extremely rich at high latitudes.

Acknowledgments: The authors are listed alphabetically as all contributed equally to the production of this work. The second author wishes to express his thanks to the staff of the Royal Tyrrell Museum for permission to collaborate in this study and for their invaluable assistance in all aspects of his work. Field assistants from the Tyrrell Museum included Kevin Lyseng and Sue Marsland. Dr. Robert "Bert" Hunt, sometimes in conjunction with Dr. Desh Mittra, of Grande Prairie Regional College, spent many hours working at the site and arranged for some of the volunteers, equipment, and the use of facilities. The number of volunteers who assisted us in the field over the years is impressively long, so we cannot list them all individually here. However, it is worth noting that two of their number (Dr. Hans C.E. Larsson of McGill University and Dr. David Varricchio of the Museum of the Rockies) started their field careers in the Pipestone Creek bone bed. We hope that all of the volunteers will take some pride in the role they played in this research. We are grateful to Dr. David Eberth (Royal Tyrrell Museum of Palaeontology) for taking the time to visit the site and for analyzing samples (with A. Deino at the University of California, Berkeley) for radiometric dating. Palynological sample processing was done by Russ Harms (Global Geolab Ltd., Medicine Hat, Alberta) with financial support from the Grande Prairie Regional College. The follow-up palynological analysis and work was done by Dr. Dennis Braman (Royal Tyrrell Museum of Palaeontology), who graciously gave us access to his unpublished reports. Rod Morgan (Calgary) is to be commended for the hundreds of volunteer hours he put into the specimen drawings. Additional artwork by Michael Skrepnick (Okatoks, cover, and Fig. 4B), Kent Wallis (Calgary, Fig. 5 crocodile scute), and the first author (Figs. 8, 12, 14, 16, 19, 20, 22, 23, 31, 37, 38, 39, 41, 46, 47, 48) supplemented Rod's work. Donna Sloan (Tyrrell Museum) produced the quarry drawing (Fig. 3) from the original quarry maps done by the third author, and the shading was done by Tetsuto Miyasita (University of Alberta). Donna Sloan, Clive Coy, and Tetsuto Miyasita also assisted in tweeking other drawings for publication. Dr. Ray Rogers did some of the thin sections of bone and rock. Assistance with the histological preparations by Dr. John S. Buckley, Greg Thompson, and David Hicks is gratefully acknowledged. Parts of this manuscript have been reviewed by Dr. Scott Sampson (University of Utah, Salt Lake City) and Dr. Michael Ryan (Cleveland Museum of Natural History), who also provided advice, encouragement, and information stemming from their own research. The original stratigraphic column (Fig. 2) was done by Gopal Dongol and was redrafted by Dave Eberth. Dr. James Gardner and Jackie Wilke of the Collections Department of the Tyrrell Museum have provided invaluable help in finding and providing data and specimens associated with this massive collection. Dr. Eva Koppelhus (University of Alberta) went well above and beyond the call of duty in handling the logistics to keep this massive project on track. Detailed reviews by Dr. Dale Russell and an anonymous reviewer led to many improvements in the final draft of the manuscript. Special thanks to Jordan Mallon (University of Calgary) and George Olshevsky (San Diego) for correcting the spelling of the species designation.

References

Acorn, J. 1993. Dinosaur world tour, official souvenir catalogue. Ex Terra Foundation, Edmonton, Alta.

Andersen, S.R.G. 2000. Studies of *Styracosaurus* from bonebed material in Dinosaur Provincial Park, Alberta, Canada: ontogeny and taphonomy. M.Sc. thesis, University of Copenhagen, Copenhagen, Denmark.

Anderson, J.S. 1999. Occipital condyle in the ceratopsian dinosaur *Triceratops*, with comments on body size variation. Museum of Paleontology, University of Michigan, Ann Arbor, Mich. Contributions 30, pp. 215–231.

Brown, B. 1914*a*. *Anchiceratops*, a new genus of horned dinosaur from the Edmonton Cretaceous of Alberta. With discussion of the origin of the ceratopsian crest and the brain casts of Anchiceratops and *Trachodon*. American Museum of Natural History Bulletin, **33**: 539–548.

Brown, B. 1914*b*. A complete skull of *Monoclonius*, from the Belly River Cretaceous of Alberta. American Museum of Natural History Bulletin, **33**: 549–558.

Brown, B., and Schlaikjer, E.M. 1940*a*. The structure and relationships of *Protoceratops*. Annals of the New York Academy of Sciences, **40**: 133–266.

Chinnery, B.J., and Weishampel, D B. 1998. *Montanoceratops cerorhynchus* (Dinosauria: Ceratopsia) and relationships among basal neoceratopsians. Journal of Vertebrate Paleontology, **18**: 569–585.

Clemens, W.A., and Nelms, L.G. 1993. Paleoecological implications of Alaskan terrestrial vertebrate fauna in latest Cretaceous time at high paleolatitudes. Geology, **21**: 503–506.

Colbert, E.H. 1945. The hyoid bones in *Protoceratops* and *Psittacosaurus*. American Museum Novitates, **1301**: 1–10.

Coombs, W.P., Jr. 1972. The bony eyelid of *Euoplocephalus* (Reptilia, Ornithischia). Journal of Paleontology, **46**: 637–650.

Currie, P.J. 1981. Hunting dinosaurs in Alberta's huge bone-bed. Canadian Geographic, **101**(4): 32–39.
Currie, P.J. 1989. Long distance dinosaurs. Natural History, **6**: 60–65.
Currie, P.J. 1992. Migrating dinosaurs. *In* The ultimate dinosaur. *Edited by* B. Preiss and R. Silverberg. Bantam Books, New York. pp. 183–195.
Currie, P.J., and Dodson, P. 1984. Mass death of a herd of ceratopsian dinosaurs. *In* Third Symposium on Mesozoic Terrestrial Ecosystems, Short Papers. *Edited by* W.-E. Reif and F. Westphal. Attempto Verlag, Tübingen. pp. 61–66.
Dawson, F.M., Jerzykiewicz, T., and Sweet, A.R. 1989. A preliminary analysis of the stratigraphy and sedimentology of the coal bearing Wapiti Group, northwestern Alberta. *In* Western Canadian Coal Geoscience—Forum Proceedings, Edmonton, Alta., 24–25 April 1989. *Edited by* W. Langenber. Alberta Research Council, Edmonton, Alta. Advances in Information Series 10, pp. 1–11.
Dodson, P. 1976. Quantitative aspects of relative growth and sexual dimorphism in *Protoceratops*. Journal of Paleontology, **50**: 929–940.
Dodson, P. 1984. Small Judithian ceratopsids, Montana and Alberta. *In* Third Symposium on Mesozoic Terrestrial Ecosystems, Short Papers, Tübingen, Germany. *Edited by* W.-E. Reif and F. Westphal. Attempto Verlag, Tübingen, Germany. pp. 73–78.
Dodson, P. 1986. *Avaceratops lammersi*: a new ceratopsid from the Judith River Formation of Montana. Academy of Natural Sciences of Philadelphia, Proceedings, **138**: 305–317.
Dodson, P. 1990. On the status of the ceratopsids *Monoclonius* and *Centrosaurus*. *In* Dinosaur systematics: perspectives and approaches. *Edited by* K. Carpenter and P.J. Currie. Cambridge University Press, Cambridge, UK. pp. 231–243.
Dodson, P. 1996. Horned dinosaurs. Princeton University Press, Princeton, N.J.
Dodson, P., and Currie, P.J. 1988. The smallest ceratopsid skull—Judith River Formation of Alberta. Canadian Journal of Earth Sciences, **25**: 926–930.
Dodson, P., and Currie, P.J. 1990. Neoceratopsia. *In* The Dinosauria. *Edited by* D.B. Weishampel, P. Dodson, and H. Osmólska. University of California Press, Los Angeles, Calif. pp. 593–618.
Dodson, P., Forster, C.A,. and Sampson, S.D. 2004. Ceratopsidae. *In* The Dinosauria. *Edited by* D.B. Weishampel, P. Dodson, and H. Osmólska. University of California Press, Berkeley, Calif. pp. 494–513.

Dong, Z.M., and Currie, P.J. 1993. Protoceratopsian embryos from Inner Mongolia, China. Canadian Journal of Earth Sciences, **30**: 2248–2254.

Eberth, D.A. 2005. The geology. *In* Dinosaur Provincial Park, a spectacular ancient ecosystem revealed. *Edited by* P.J. Currie and E.B. Koppelhus. Indiana University Press, Bloomington, Ind. pp. 54–82.

Eberth, D.A., and Getty, M.A. 2005. Ceratopsian bonebeds: occurrence, origins and significance. *In* Dinosaur Provincial Park, a spectacular ancient ecosystem revealed. *Edited by* P.J. Currie and E.B. Koppelhus. Indiana University Press, Bloomington, Ind. pp. 501–536.

Eberth, D.A., Braman, D.R., and Tokaryk, T.T. 1990. Stratigraphy, sedimentology and vertebrate palaeontology of the Judith River Formation (Campanian) near Muddy Lake, west-central Saskatchewan. Bulletin of Canadian Petroleum Geology, **38**: 387–406.

Fanti, F., and Currie, P.J. 2007. A new *Pachyrhinosaurus* bonebed from the late Cretaceous Wapiti Formation. Ceratopsian Symposium 2007, Royal Tyrrell Museum, Short Papers, Abstracts and Programs, pp. 39–43.

Farke, A.A. 2004. Horn use in *Triceratops* (Dinosauria: Ceratopsidae): testing behavioral hypotheses using scale models [online]. Palaeontologia Electronica, 7(1). Available from http://palaeo-electronica.org/2004_1/horn/issue1_04.htm.

Farke, A.A., and Williamson, T. 2005. A chasmosaurine ceratopsid parietal from the Naashoibito Member, Oho Alamo Formation of New Mexico, with implications for ceratopsid systematics and biogeography. Journal of Vertebrate Paleontology, **25**(Suppl.): 55A. [Abstr.]

Farlow, J.O., and Dodson, P. 1975. The behavioral significance of the frill and horn morphology in ceratopsian dinosaurs. Evolution, **29**: 353–361.

Fiorillo, A.R. 1991. Taphonomy and depositional setting of Careless Creek Quarry (Judith River Formation), Wheatland County, Montana, USA. Palaeogeography, Palaeoclimatology, Palaeoecology, **81**: 281–311.

Fiorillo, A.R. 2004. The dinosaurs of Arctic Alaska. Scientific American, **291**(6): 84–91.

Fiorillo, A.R., and Gangloff, R. 2003. Preliminary notes on the taphonomic and paleoecologic setting of a *Pachyrhinosaurus* bonebed in northern Alaska. Journal of Vertebrate Paleontology, **23**(Suppl.): 50A. [Abstr.]

Forster, C.A. 1990. The cranial morphology and systematics of *Triceratops* with a preliminary analysis of ceratopsian phylogeny. Ph.D. thesis, Department of Geology, University of Pennsylvania, Philadelphia, Pa.

Forster, C.A. 1996*a*. New information on the skull of *Triceratops*. Journal of Vertebrate Paleontology, **16**: 246–258.

Forster, C.A. 1996*a*. Species resolution in *Triceratops*: cladistic and morphometric approaches. Journal of Vertebrate Paleontology, **16**: 259–270.

Forster, C.A., Sereno, P.C., Evans, T.W., and Rowe, T. 1993. A complete skull of *Chasmosaurus mariscalensis* (Dinosauria: Ceratosidae) from the Aguja Formation (late Campanian) of west Texas. Journal of Vertebrate Paleontology, **13**: 161–170.

Gilmore, C.W. 1917. *Brachyceratops*, a ceratopsian dinosaur from the Two Medicine of Montana, with notes on associated fossil reptiles. United States Geological Survey Professional Paper, 53. pp. 1–45.

Gilmore, C.W. 1946. Reptilian fauna of the North Horn Formation of central Utah. Part 2. Description of a new species of Ceratopsia. United States Geological Survey Professional Paper, 210-C.

Goodwin, M.B., Clemens, W.A., Horner, J.R., and Padian, K. 2006. The smallest known *Triceratops* skull: new observations on ceratopsid cranial anatomy and ontogeny. Journal of Vertebrate Paleontology, **26**: 103–112.

Granger, W., and Gregory, W.K. 1923. *Protoceratops andrewsi*, a pre-ceratopsian dinosaur from Mongolia. American Museum Novitates, **72**: 1–9.

Haas, G. 1955. The jaw musculature in *Protoceratops* and in other ceratopsians. American Museum Novitates, **1729**: 1–24.

Hamada, Y. (*Editor*). 2001. Dinosaurs of the Royal Tyrrell Museum of Palaeontology. Fukui Prefectural Dinosaur Museum, Katsuyama, Japan.

Hatcher, J.B., Marsh, O.C., and Lull, R.S. 1907. The Ceratopsia. U.S. Geological Survey Monograph, **49**: 1–198.

Hieronymus, T., Tanke, D., Currie, P.J., and Witmer, L. 2006. Horn morphology of *Pachyrhinosaurus* and horn evolution in centrosaurine dinosaurs. Journal of Vertebrate Paleontology, **26**: 75A. [Abstr.]

Holmes, R.B., Forster, C., Ryan, M., and Shepherd, K.M. 2001. A new species of *Chasmosaurus* (Dinosauria: Ceratopsia) from the Dinosaur Park Formation of southern Alberta. Canadian Journal of Earth Sciences, **38**: 1423–1438.

Hopson, J.A. 1979. Paleoneurology. *In* Biology of the Reptilia. Vol. 9. Neurology A. *Edited by* C. Gans. Academic Press, London. pp. 39–146.

Horner, J. R., and Currie, P. J. 1994. Embryonic and neonatal morphology and ontogeny of a new species of *Hypacrosaurus* (Ornithischia, Lambeosauridae) from Montana and Alberta. *In* Dinosaur eggs and babies. *Edited by* K. Carpenter, K.F. Hirsch, and J.R. Horner. Cambridge University Press, Cambridge, UK. pp. 312–336.

Horner, J.R., and Goodwin, M.B. 2006. Major cranial changes during *Triceratops* ontogeny. Proceedings of the Royal Society B, **273**: 2757–2761.

Hotton, N., III. 1980. An alternative to dinosaur endothermy, the happy wanderers. *In* A cold look at the warm-blooded dinosaurs. *Edited by* R.D.K. Thomas and E.C. Olson. Westview Press, Boulder, Colo. pp. 311–350.

Kurzanov, S. M. 1972. Sexual dimorphism in protoceratopsians. Paleontological Journal, 1972: 91–97.

Lambe, L.M. 1915. On *Eoceratops canadensis*, gen. nov., with remarks on other genera of Cretaceous horned dinosaurs. Canadian Department of Mines, Geological Survey Museum Bulletin, **12**: 1–49.

Langston, W., Jr. 1960. The vertebrate fauna of the Selma Formation of Alabama. Part VI. The dinosaurs. Fieldiana: Geology Memoirs, **3**: 319–361.

Langston, W., Jr. 1967. The thick-headed ceratopsian dinosaur *Pachyrhinosaurus* (Reptilia: Ornithischia) from the Edmonton Formation near Drumheller, Canada. Canadian Journal of Earth Sciences, **4**: 171–186.

Langston, W., Jr. 1968. A further note on *Pachyrhinosaurus* (Reptilia: Ceratopsia). Journal of Paleontology, **42**: 1303–1304.

Langston, W., Jr. 1975. The ceratopsian dinosaurs and associated lower vertebrates from the St. Mary River Formation (Maestrichtian) at Scabby Butte, southern Alberta. Canadian Journal of Earth Sciences, **12**: 1576–1608.

Langston, W., Jr. 1976. A Late Cretaceous vertebrate fauna from the St. Mary River Formation in western Canada. *In* Athlon, essays on palaeontology in honour of Loris Shano Russell. *Edited by* C.S. Churcher. Royal Ontario Museum, Toronto, Ont. pp. 114–133.

Lehman, T.M. 1989. *Chasmosaurus mariscalensis* sp. nov., a new ceratopsian dinosaur from Texas. Journal of Vertebrate Paleontology, **9**: 137–162.

Lehman, T.M. 1990. The ceratopsian subfamily Chasmosaurinae: sexual dimorphism and systematics. *In* Dinosaur systematics: perspectives and approaches. *Edited by* K. Carpenter and P.J. Currie. Cambridge University Press, Cambridge, UK. pp. 211–229.

Lehman, T.M. 1996. A horned dinosaur from the El Picachu Formation of west Texas and review of ceratopsian dinosaurs from the American southwest. Journal of Paleontology, **70**: 494–508.

Lull, R.S. 1908. The cranial musculature and the origin of the frill in the ceratopsian dinosaurs. American Journal of Science, **25**: 387–398.

Lull, R.S. 1933. A revision of the Ceratopsia. Peabody Museum of Natural History, Memoir, **3**: 1–175.

Maddison, D.R., and Maddison, W.P. 2005. MacClade 4: analysis of phylogeny and character evolution, version 4.08. Sinauer Associates, Sunderland, Mass.

Makovicky, P.J. 2002. Taxonomic revision and phylogenetic relationships of basal Neoceratopsia (Dinosauria: Ornithischia). Ph.D. thesis, Columbia University, New York.

Maryanska, T., and Osmólska, H. 1975. Protoceratopsidae (Dinosauria) of Asia. Palaeontologia Polonica, **33**: 133–182.

May, K.C., and Gangloff, R.A. 1999. New dinosaur bone bed from the Prince Creek Formation, Colville River, National Petroleum Reserve-Alaska. Journal of Vertebrate Paleontology, **19**(Suppl.): 62A. [Abstr.]

Nelms, L.G., and Clemens, W.A. 1989. *Pachyrhinosaurus* from the Kogosuhruk Tongue, Prince Creek Formation (Late Cretaceous), north slope of Alaska. American Association for the Advancement of Science, 40th Arctic Science Conference, Fairbanks, Alaska. Proceedings, p. 36.

Ostrom, J.H. 1966. Functional morphology and evolution of the ceratopsian dinosaurs. Evolution, **20**: 290–308.

Ostrom, J.H., and Wellnhofer, P. 1986. The Munich specimen of *Triceratops* with a revision of the genus. Zitteliana, **14**: 111–158.

Parks, W.A. 1921. The head and fore limb of a specimen of *Centrosaurus apertus*. Transactions of the Royal Society of Canada, Section IV, **15**: 53–62.

Penkalski, P., and Dodson, P. 1999. The morphology and systematics of *Avaceratops*, a primitive horned dinosaur from the Judith River Formation (Late Campanian) of Montana, with the description of a second skull. Journal of Vertebrate Paleontology, **19**: 692–711.

Rogers, R. 1990. Taphonomy of three dinosaur bone beds in the Upper Cretaceous Two Medicine Formation of northwestern Montana: evidence for drought-related mortality. Palaios, **5**: 394–413.

Rogers, R.R., Eberth, D.A., and Fiorillo, A.R. 2007. Bonebeds: genesis, analysis and paleobiological significance. University of Chicago Press, Chicago, Ill.

Romer, A.S. 1956. Osteology of the reptiles. University of Chicago Press, Chicago, Ill.
Rothschild, B.M., and Tanke, D. 1992. Paleopathology of vertebrates: insights to lifestyle and health in the geological record. Geoscience Canada, **19**(2): 73–82.
Rothschild, B.M., and Tanke, D. 1996. Thunder in the Cretaceous: interspecies conflict as evidence for ceratopsian migration. *In* Dinofest International Symposium. Program and Abstracts. *Edited by* D.L. Wolberg and E. Stump. April 18–21, 1996. Arizona State University, Tempe, Ariz. p. 96. [Abstr.]
Rothschild, B.M., and Tanke, D. 1997. Thunder in the Cretaceous: interspecies conflict as evidence for ceratopsian migration? *In* Dinofest International. Proceedings of a Symposium Sponsored by Arizona State University. *Edited by* D.L. Wolberg, E. Stump, and G.D. Rosenberg. pp. 77–82.
Russell, D. A. 1984. A check list of the families and genera of North American dinosaurs. Syllogeus, **53**: 1–35.
Ryan, M.J. 1992. The taphonomy of a *Centrosaurus* (Reptilia: Ornithischia) bone bed (Campanian), Dinosaur Provincial Park, Alberta, Canada. M.Sc. thesis, University of Calgary, Calgary, Alta.
Ryan, M.J. 1994. Taphonomy of a *Centrosaurus* (Ornithischia: Ceratopsidae) bonebed from the Dinosaur Park Formation (Upper Campanian), Alberta, Canada. Journal of Vertebrate Paleontology, **14**(Suppl.): 44A. [Abstr.]
Ryan, M.J. 2003. Taxonomy, systematics and evolution of centrosaurine ceratopsids of the Campanian Western Interior Basin of North America. Ph.D. thesis, Department of Biological Sciences, University of Calgary, Calgary, Alta.
Ryan, M.J. 2007. A new basal centrosaurine ceratopsid from the Oldman Formation, southeastern Alberta. Journal of Paleontology, **81**: 376–396.
Ryan, M.J., and Russell, A.P. 2005. A new centrosaurine ceratopsid from the Oldman Formation of Alberta and its implications for centrosaurine taxonomy and systematics. Canadian Journal of Earth Sciences, **42**: 1369–1387.
Ryan, M.J., Russell, A.P., Eberth, D.A., and Currie, P.J. 2001. The taphonomy of a *Centrosaurus* (Ornithischia: Ceratopsidae) bone bed from the Dinosaur Park Formation (Upper Campanian), Alberta, Canada, with comments on cranial ontogeny. Palaios, **16**: 482–506.
Ryan, M.J., Brinkman, D., Eberth, D., Currie, P.J., and Tanke, D.H. 2006. A new *Pachyrhinosaurus*-like ceratospian from the Upper Dinosaur Park Formation (Late Campanian) of southern Alberta. Journal of Vertebrate Paleontology, **26**(Suppl): 117A. [Abstr.]

Sampson, S.D. 1993. Cranial ornamentations in ceratopside dinosaurs: systematic, behavioural and evolutionary implications. Ph.D. thesis, University of Toronto, Toronto, Ont.

Sampson, S.D. 1995. Two new horned dinosaurs from the Upper Cretaceous Two Medicine Formation of Montana, USA; with a phylogenetic analysis of the Centrosaurinae (Ornithischia: Ceratopsidae). Journal of Vertebrate Paleontology, **15**: 743–760.

Sampson, S.D. 1997*a*. Variation. *In* Encyclopedia of dinosaurs. *Edited by* P. J. Currie and K. Padian. Academic Press, San Diego, Calif. pp. 773–780.

Sampson, S.D. 1997*b*. Bizarre structures and dinosaur evolution. *In* Dinofest International. *Edited by* D.L. Wolberg, E. Stump, and G.D. Rosenberg. Academy of Natural Sciences, Philadelphia, Pa. pp. 39–45.

Sampson, S.D., and Tanke, D.H. 1990. Ontogeny and variation in the crania of centrosaurine ceratopsians (Ornithischia: Ceratopsidae): phylogenetic implications. Journal of Vertebrate Paleontology, **10**: 40A. [Abstr.]

Sampson, S.D., Ryan, M.J., and Tanke, D.H. 1997. Craniofacial ontogeny in centrosaurine dinosaurs (Ornithischia: Ceratopsidae): taxonomic and behavioral implications. Zoological Journal of the Linnean Society, **121**: 293–337.

Schweitzer, M.H., Wittmeyer, J.L., Horner, J.R., and Toporski, J.K. 2005. Soft-tissue vessels and cellular preservation in *Tyrannosaurus rex*. Science (Washington, D.C.), **307**: 1952–1955.

Sereno, P.C. 1986. Phylogeny of the bird-hipped dinosaurs. National Geographic Research, **2**: 234–236.

Sereno, P.C. 1990. Psittacosauridae. *In* The Dinosauria. *Edited by* D.B. Weishampel, P. Dodson, and H. Osmólska. University of California Press, Berkeley, Calif. pp. 579–592.

Spassov, N.B. 1979. Sexual selection and the evolution of horn-like structure of ceratopsian dinosaurs. Palaeontology, Stratigraphy and Lithology (Bulgarian Academy of Science), **11**: 37–48.

Spotila, J.R., O'Connor, M P, Dodson, P., and Paladino, F.V. 1991. Hot and cold running dinosaurs: body size, metabolism and migration. Modern Geology, **16**: 203–227.

Sternberg, C.M. 1926. Notes on the Edmonton Formation of Alberta. Canadian Field-Naturalist, **40**: 102–104.

Sternberg, C.M. 1927. Homologies of certain bones of the ceratopsian skull. Transactions of the Royal Society of Canada, Section IV, **21**: 135–143.

Sternberg, C.M. 1940. Ceratopsidae from Alberta. Journal of Paleontology, **14**: 468–480.

Sternberg, C.M. 1947. New dinosaur from southern Alberta, representing a new family of the Ceratopsia. Geological Society of America, **58**: 1230.

Sternberg, C.M. 1949. The Edmonton fauna and description of a new *Triceratops* from the Upper Edmonton Member: phylogeny of Ceratopsidae. National Museum of Canada, Bulletin, **113**: 33–46.

Sternberg, C.M. 1950. *Pachyrhinosaurus canadensis*, representing a new family of the Ceratopsia, from southern Alberta. National Museum of Canada, Bulletin, **118**: 109–120.

Sternberg, C.M. 1951. Complete skeleton of *Leptoceratops gracilis* Brown from the Upper Edmonton Member on the Red Deer River, Alberta. National Museum of Canada, Bulletin, **123**: 225–255.

Swofford, D.L. 2002. PAUP 4.0b10, beta version. Sinauer Associates, Sunderland, Mass.

Tanke, D.H. 1988. Ontogeny and dimorphism in *Pachyrhinosaurus* (Reptilia, Ceratopsidae) Pipestone Creek, N.W. Alberta, Canada. Journal of Vertebrate Paleontology, **8** (Suppl.): 27A. [Abstr.]

Tanke, D.H. 1989. K/U Centrosaurine (Ornithischia: Ceratopsidae) paleopathologies and behavioral implications. Journal of Vertebrate Paleontology, **9**(Suppl.): 41A. [Abstr.]

Tanke, D.H. 2005. The Late Cretaceous pachyrhinosaur bonebed (Late Cretaceous: Wapiti Formation) near Grande Prairie, Alberta. *In* Alberta Palaeontological Society Ninth Annual Symposium, Abstracts. *Edited by* H. Allen. Alberta Palaeontological Society, Calgary, Alta. pp. 33–35.

Tanke, D.H. 2006. Sixty years of pachyrhinosaur (Dinosauria: Ceratopsidae) discoveries in North America. *In* Alberta Palaeontological Society Tenth Annual Symposium, Abstracts. *Edited by* H. Allen. Alberta Palaeontological Society, Calgary, Alta. pp. 38–56.

Tanke, D.H., and Farke, A.A. 2003. Cranial abnormalities in horned dinosaurs: disease and normal biological processes—not combat wounds. *In* Alberta Palaeontological Society, Seventh Annual Symposium, Abstracts Volume. *Edited by* H. Allen. Alberta Palaeontological Society, Calgary, Alta. pp. 78–81.

Tanke, D.H., and Farke, A.A. 2006. Bone resorption, bone lesions and extracranial fenestrae in ceratopsid dinosaurs: a preliminary assessment. *In* Horns and beaks, ceratopsian and ornithopod dinosaurs. *Edited by* K. Carpenter. Indiana University Press, Bloomington, Ind. pp. 319–347.

Tanke, D.H., and Rothschild, B.M. 1997. Paleopathology. *In* Encyclopedia of dinosaurs. *Edited by* P.J. Currie and K. Padian. Academic Press, London. pp. 525–530.

Tanke, D.H., and Rothschild, B.M. 2002. DINOSORES: an annotated bibliography of dinosaur paleopathology and related topics—1838–2001. New Mexico Museum of Natural Hisory and. Science, Bulletin, 20.

Tyson, H. 1977. Functional craniology of the Ceratopsia (Reptilia: Ornithischia) with special reference to *Eoceratops*. M.Sc. thesis, University of Alberta, Edmonton, Alta.

Varricchio, D.J., and Horner, J.R. 1993. Hadrosaurid and lambeosaurid bone beds from the Upper Cretaceous Two Medicine Formation of Montana. Canadian Journal of Earth Sciences, **30**: 997–1006.

Visser, J. 1986. Sedimentology and taphonomy of a *Styracosaurus* bonebed in the Late Cretaceous Judith River Formation, Dinosaur Provincial Park, Alberta. M.Sc. thesis, Department of Geology and Geophysics, University of Calgary, Calgary, Alta.

von Huene, F. 1950. Bermerkungen zu einem fremdartigen neuen Ceratopsiden. Neues Jahrbuch für Geologie und Paläontologie Monatshefte, **11**: 347–351.

Wolfe, D.G. 2000. New information on the skull of *Zuniceratops christopheri*, a neoceratopsian dinosaur from the Cretaceous Moreno Hill Formation, New Mexico. New Mexico Museum of Natural History and Science Bulletin, **17**: 93–94.

Wolfe, D.G., and Kirkland, J.I. 1998. *Zuniceratops christopheri* n. gen., and sp., a ceratopsian dinosaur from the Moreno Hill Formation (Cretaceous, Turonian) of west-central New Mexico. New Mexico Museum of Natural History and Science Bulletin, **14**: 303–317

Wolfe, D.G., Beekman, S., Mcguiness, D., Robira, T., and Denton, R. 2004. Taphonomic characterization of a *Zuniceratops* bone bed from the middle Cretaceous (Turonian) Moreno Hill Formation. Journal of Vertebrate Paleontology, **24** (Suppl.): 131A. [Abstr.]

Wood, J.M., Thomas, R.G., and Visser, J. 1988. Fluvial processes and vertebrate taphonomy: the Upper Cretaceous Judith River Formation, south-central Dinosaur Provincial Park, Alberta, Canada. Palaeogeography, Palaeoclimatology, Palaeoecology, **66**: 127–143.

You H.-L., and Dodson, P. 2004. Basal ceratopsia. *In* The Dinosauria. *Edited by* D.B. Weisampel, P. Dodson, and H. Osmólska. University of California Press, Berkeley, Calif. pp. 478–493.

Appendix A

List of characters and character states used to resolve relationships of seven taxa within the Centrosaurinae.

Morphological characters used in this paper, most of which are derived from Sampson (1995) and Ryan (2007), are unordered and unweighted. The basal state is assigned a value of zero. References are included for authors who introduced, discussed, or modified the specific characters.

1. Rostral, size and shape: triangular in lateral view, with short dorsal and ventral processes (0); elongate, with deeply concave posterior margin and hypertrophied dorsal and ventral processes (1) (Sereno 1986).
2. Premaxilla, septum: absent (0); present and subcircular (1); present and anteriorly elongate (2) (Ryan 2007).
3. Premaxilla, septum within narial chamber: absent (0); thick and has simple, plate-like construction (1); thin, often with transverse perforations (2) (Langston 1967; Forster 1996*b*; Holmes et al. 2001).
4. Premaxilla, premaxillary (narial) process extending into the external naris from the caudoventral margin of the premaxillary septum: absent (0); present (1) (Forster 1990).
5. Premaxilla, thickened narial strut (separating fenestra through septum from narial opening) along posterior border of premaxillary septum: absent (0); present, anteriorly inclined (1); present, caudally inclined (2) (Forester et al. 1993; Holmes et al. 2001; Ryan 2007).
6. Premaxilla, ventral expansion of posteroventral margin: absent, posteroventral margin of premaxilla unexpanded and level with alveolar margin of maxilla (0); present, expanded ventrally to extend well below alveolar margin of the maxilla (1) (Sereno 1986; Penkalski and Dodson 1999).
7. Premaxilla, posterior tip of posteroventral process inserts into embayment in the nasal and is surrounded by the nasal: yes (0); no (1) (Forster et al. 1993, Holmes et al. 2001).
8. Premaxilla contact with lacrimal: separated by nasal and maxilla (0); in contact (1) (Lull 1933; Makovicky 2002).
9. External antorbital fenestra, size: large, 20% or more length of body of maxilla (0); greatly reduced to less than 10% length of body of maxilla (1); reduced to foramen or absent (2) (Granger and Gregory 1923; Chinnery and Weishampel 1998).
10. Nasal, basal length of horn or boss (subadult): short based, restricted in length anteroposteriorly (0); long based, ornamentation covers almost the entire length of the nasal (1) (Sampson et al. 1997).

11. Nasal, ornamentation type (adult): absent or poorly developed, less than 15% basal skull length (0); elongated horn, more than 20% basal skull length (1); long based, low thickened ridge (2); boss (3) (Forester et al. 1993; Sampson 1995; Ryan 2007).
12. Nasal, posterior margin of external naris: concave (0); pronounced tab-like process projecting anteriorly into the nasal vestibule (1) (Langston 1975; Sereno 1986).
13. Jugal, infratemporal flange (adult): absent (0); present, contacts jugal flange of squamosal under infratemporal fenestra (1) (Brown and Schlaikjer 1940; Lehman 1996; Forster 1996*b*; Ryan 2007).
14. Prefrontal: separated by frontals and excluded from margins of frontal fontanelle (0); contact each other on midline, separate nasals from frontals and form anterior margin of frontal fontanelle (1) (Lambe 1915; Forster 1990; Ryan 2003).
15. Prefrontal and lacrimal: form prominent antorbital buttress (0); do not form antorbital buttress (1).
16. Postorbital, form of postorbital ornamentation (subadult): conical horn core (height at least 3 times anteroposterior basal length) with rounded base and pointed apex (0); pyramidal horn core (approximately 1:1 ratio of height to anteroposterior basal length) (1); horn core longer anteroposteriorly than high, with rounded apex (2) (Sampson 1995).
17. Supraorbital ornamentation type (adult): absent (0); present, horn (1) ; present, boss (2) (Sampson 1995).
18. Postorbital, horn core shape (unmodified adult): elongate with pointed apex and round to oval base (0); pyramidal with rounded apex, at least as tall as base is long (1); rounded apex, base longer than horn is tall (2) (Sampson 1995; Ryan 2007).
19. Postorbital, horn core height (unmodified adult): long, greater than 60% length of face (0); short, less than 40% length of face (1) (Forster 1990; Holmes et al. 2001; Ryan 2007).
20. Postorbital, position of base of horn core (adult): posterior to orbit (0); over or anterior to orbit (1) (Lehman 1996).
21. Postorbital horn core curvature (adult): straight, dorsally, anteriorly or anterolaterally curved (0); curves posteriorly (1) (Forster et al. 1993; Lehman 1996; Holmes et al. 2001).
22. Parietosquamosal frill, length relative to basal skull length: elongate, 0.80 or more (0); shortened, 0.70 or less (1) (Hatcher et al. 1907; Lehman 1990).
23. Squamosal, length of squamosal relative to parietal: equal or subequal in length (0); squamosal less than 60% total parietal length (1) (Sereno 1986).

24. Squamosal, shape of posterodorsal (medial) margin of squamosal: straight (0); posterior portion stepped-up relative to anterior portion, transition taking place where quadrate groove passes from ventral to dorsal surface of the bone (1) (Dodson 1986; Penkalski and Dodson 1999; Ryan 2007).
25. Squamosal, anteromedial lamina forming the posterolateral floor of dorsotemporal fossa: absent (0); present (1) (Dodson 1986).
26. Parietal, dorsal surface of medial bar: smooth and flat (chasmosaurines) (0); small, rounded midline humps (centrosaurines) (1); large spikes (2) (Sampson 1995).
27. Parietal, posterior surface on midline: posteriorly convex or straight (*Protoceratops* and most chasmosaurines) (0); deep, U-shaped emargination (1) (centrosaurines plus *Chasmosaurus*).
28. Epoccipital, profile shape of epoccipitals on squamosal: not present (0); crescentic to lozenge shaped (1); triangular (2) (Holmes et al. 2001).
29. Epoccipital, number of loci for epoccipitals on parietal rami lateral to the midline margin: none (0); three to five (1); six to eight (2) (Ryan 2007).
30. Epoccipital, most medial process (locus 1): absent (0); unelaborated epoccipital on posterior margin (1); short (length of hook equal to diameter of base) procurving hook on dorsal margin (2); long procurving hook on dorsal margin with length of hook twice the diameter of base (3); triangular epoccipital on dorsal margin (4) (Sampson 1995; Ryan 2007).
31. Epoccipital, orientation of most medial epoccipital (locus 1): absent (0); posteriorly directed (1); dorsally directed (2); anteriorly directed in pronounced anterior curl (3) (Sampson 1995; Ryan 2007).
32. Epoccipital, process at locus 2: absent (0); unelaborated epoccipital on posterior margin (1); small, medially curled hook (length of hook less than or equal to length of base) (2); large, medially curled hook (length of hook at least twice length of base) (3); multipronged, posteriorly directed process (4); large, triangular profile (5) (Sampson 1995; Ryan and Russell 2005; Ryan 2007).
33. Epoccipital, process at locus 3: absent (0); small, unelaborated epoccipital on posterior margin (1); narrow-based hook (length between 1 and 3 times basal diameter) (2); narrow-based long spike greater than 4 times basal diameter (3); broad-based short pachyostotic spike (4) (Sampson 1995; Ryan and Russell 2005; Ryan 2007).
34. Epoccipital, orientation of spike-like epoccipital at locus 3: absent or small (0); posteriorly directed (1); posterolaterally directed (2); laterally or anterolaterally directed (3); dorsolaterally directed (4) (Sampson 1995; Ryan 2007).

35. Epoccipitals, pattern of fusion to frill margin: occurs from rostral to caudal (0); occurs from caudal to rostral (1) (Lehman 1990).
36. Parietal processes at loci 4–6: absent (0); small (1); large spikes (2) (Sampson 1995).
37. Epoccipital, imbrication of lateral marginal undulations of the parietal: absent (0); present (1) (Sampson et al. 1997).
38. Predentary, orientation of the triturating surface relative to the horizontal plane of the element: nearly horizontal (0); steeply inclined laterally (1) (Lehman 1990; Forster 1996*b*).
39. Dentary, coronoid process: incipient process with gently convex apex and no neck (0); well developed but lacks anterior extension distally (1); high, powerful and expands anteriorly at the distal (dorsal) end (2) (Lull 1933).
40. Dentary, posterior extent of tooth row: terminates medial to coronoid process (0); terminates posterior to the coronoid process (1) (Brown and Schlaikjer 1940).
41. Teeth, roots: single (0); double (1) (Brown and Schlaikjer 1940).
42. Teeth, number of vertically stacked replacement teeth per tooth family: one or two replacement teeth (0); more than two replacement teeth (1) (Brown and Schlaikjer 1940).
43. Tooth ornamentation: subsidiary ridges present, extending from margin to base of tooth (0); subsidiary ridges reduced, present only at margin of teeth (1) (Dodson et al. 2004).
44. Sacrum, deep longitudinal channel on ventral surface: present (0); absent (1) (Lambe 1915; Lehman 1990).
45. Ischium, cross-sectional shape of shaft: ovoid (0); laterally compressed and blade-like, narrow along dorsal margin (1) (Dodson et al. 2004).
46. Ischium, orientation of shaft: nearly straight (0); slightly decurved (1); broadly and continuously curved (2) (Brown and Schlaikjer 1940).

2. Comments on the quarry map and preliminary taphonomic observations of the *Pachyrhinosaurus* (Dinosauria: Ceratopsidae) bone bed at Pipestone Creek, Alberta, Canada

PATRICIA E. RALRICK AND
DARREN H. TANKE

Abstract

The *Pachyrhinosaurus lakustai* quarry on Pipestone Creek near Grande Prairie, Alberta, Canada, is a monodominant bone bed with well-preserved fossil bones that are predominantly completely disarticulated. Nearly every element within the skeleton is represented. Identified bones were grouped by transportability of the bones present using a previously published classification system. Just over 70% of the bones were classified within the most transportable group, indicating a fluvial allochthonous accumulation. Orientations produced by utilizing the bone bed maps indicate that there is no single directionality to the bones.

Résumé

La carrière de *Pachyrhinosaurus lakustai* du ruisseau Pipestone, près de Grande Prairie (Alberta, Canada), est un dépôt ossifère dominé exclusivement par des os fossilisés très bien préservés et presque entièrement désarticulés. La quasi-totalité des éléments du squelette y est représentée. Les os recensés ont été groupés en fonction de leur transportabilité à l'aide d'un système de classification déjà publié. Un peu plus de 70 % des os ont été classés dans le groupe le plus transportable, ce qui suggère un dépôt fluvial d'origine allochtone. Les patrons d'accumulations établies à l'aide des cartes des dépôts ossifères ne révèlent aucune directionalité précise.

Introduction

The rich *Pachyrhinosaurus lakustai* bone bed on Pipestone Creek near Grande Prairie, Alberta, Canada, was systematically excavated during the summers of 1986–1989 by staff and volunteers of the Royal Tyrrell Museum of Palaeontology, Drumheller, Alberta. It was found to be an over 99% monodominant assemblage (Tanke 2004). The fossilized bones are well preserved, although many elements are crushed, and the skeletons are completely disarticulated (except some skulls and skull sections) with nearly all elements represented (Tanke 2005). The mean bone bed depth was 23–25 cm with as many as 200 elements/m^2 (Tanke 2006; Currie et al. 2008). It has been assumed to be a mass mortality event accumulated by fluvial action. To confirm the fluvial aspect of this assumption, the bone bed map was assembled, and bone orientations were calculated. The collected bones were placed within Voorhies groups (Voorhies 1969) to assess transportability and paleostream flow direction.

Mapping methods

During the initial excavation (summer of 1986), a string baseline was set up roughly parallel to the eroded edge of the outcrop closest to Pipestone Creek. The baseline was marked at 1 m intervals. Areas of uncovered bone bed were mapped using a square aluminum frame (referred to as a gridbox) with 1 m × 1 m internal dimensions and 10 cm gradations marked by string. Adjustable legs and a bubble level ensured the gridbox was level. The areas being mapped were delineated by matching the internal corners of one side of the gridbox with the 1 m gradations on the baseline. A plumb bob on the end of a string was then held vertically within each unmarked corner of the gridbox, and the spot was marked by coloured flagging tape wrapped around the top of a nail that was hammered into the rock. When a fossil coincided with a corner, it was marked with a dot of florescent orange spray paint or white typewriter correction fluid.

As excavation proceeded and the quarry surface became lower, the nails were pounded downward to each new level. Paint dots were reapplied as needed or replaced by nails. To maximize accuracy, the positions of the nails and paint dots were frequently checked against the 1 m marks on the baseline.

Because of the density of bone, the muddy nature of the site (because of springs and rainfall), repeated vandalism, and large vertebrates (such as bears) that occasionally wandered through the quarry and damaged specimens, the strike, dip, and orientation of the bones were generally not recorded because this would have slowed down the removal of specimens. Fossils were removed as quickly as possible after they were uncovered and mapped. Vandalism was particularly troublesome during the summers of 1987–1989, when equipment was stolen or damaged and exposed fossils were broken or destroyed whenever the crew was away from the quarry. Depredations by vandals proved to be so problematic in 1989 that the crew worked on site each day until 2230 h. To distract potential vandals, an artificial dig site (devoid of fossils) was set up at another outcrop several hundred metres downstream from the actual quarry.

An exploratory quarry, roughly 2.5 m × 6 m, was excavated during the summer of 1986. In the fall of that year, heavy equipment opened up a much larger pit measuring some 240 m^2 (Currie et al. 2008). The heavy equipment destroyed the walls and baseline posts of the 1986 quarry, which therefore lost its spatial context within the larger pit. Although the precise position of the 1986 section cannot be determined, the baseline is more or less parallel to that of the larger mapped section, so the bone orientations for that section should be more or less correct. Baseline posts were removed without authorization between the 1988 and 1989 field seasons, but fortunately, the 1989 maps could be joined to the 1987–1988 map sections because in situ mapping nails from 1988 were relocated.

The quarry mapping was done primarily by the second author, assisted by Museum staff and trained postsecondary students. The bone bed maps, each covering 1 m^2, were numbered with the intention of assembling them into a complete quarry map. Although part of this was done when the quarry was still in operation (Currie et al. 2008), it was not until 20 years later that the numerous map pages were reassembled by the first author. The bone outlines were traced from the grid paper to tracing paper, which was then photocopied and scanned. The sections of the bone bed map where then reassembled within Adobe Photoshop® 9.2 (Fig. 1).

Orientations for 1188 bones were determined by drawing straight lines along the long axis of bone outlines on the composite map. The orientations of elongate bones were calculated in relation to the north declination of the baseline and then were uploaded into the circular statistics software package Oriana® 2.0 to construct the rose diagram (Fig. 2).

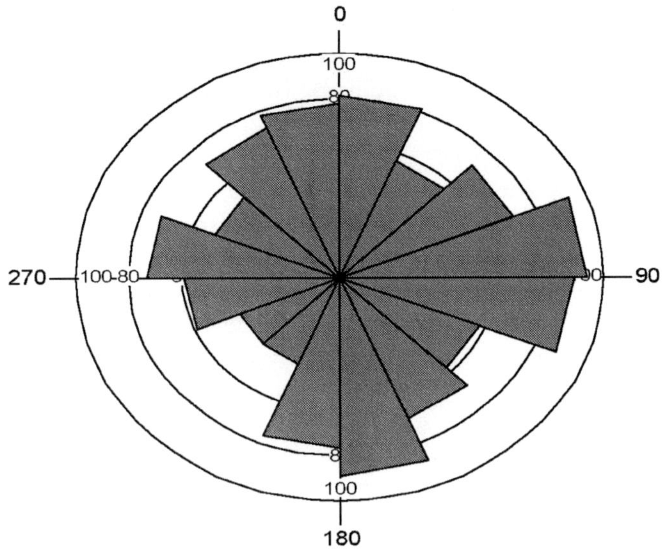

Fig. 2. Rose diagram indicating variable orientation of bones in the bone bed of Pipestone Creek (N = 1093). North is 0°, and the petal wedges are 22.5° in width.

Results and discussion

Disarticulated bones can accumulate at sites in many different ways, including fluvial or eolian processes and carnivore activity (Behrensmeyer 1975). The large size of many of the bones (mainly skulls) and study of the sediments associated with the bones clearly shows that eolian processes had nothing to do with this accumulation of *Pachyrhinosaurus* bones. The fact that few of the bones are toothmarked (Currie et al. 2008), plus the sheer quantity of bone, negates the possibility that this is a bone bed caused by carnivore activity. Fluvial processes are evident throughout the surrounding sediments, and some fish, turtle, and crocodile fossils were also recovered within the bone bed.

Because bones move within nonviscous flows at different rates, the type of elements within a site can suggest whether a bone bed is an allochthonous (transported from the place of death) or autochthonous (within the place of death) assemblage (Behrensmeyer 1975). To determine the transportability of disarticulated skeletal elements within a fluvial environment, Voorhies (1969) conducted flume experiments with coyote (*Canis latrans*) and sheep (*Ovis* sp.) skeletal elements. He found that the same elements were moved consistently within the flume and was able to separate the bones into three groups from most to least transportable.

The online records (http://hermis.cd.gov.ab.ca/rtmp/) of the Royal Tyrrell Museum of Palaeontology were accessed, and 1395 identified *P. lakustai* bones were classified into Voorhies groups (Table 1). However, a problem arose during this exercise, because the Pipestone Creek bone bed produced many more disarticulated skull elements than partial or complete skulls. Unfortunately, disarticulated skull elements were not included within Voorhies' (1969) flume experiment. Disarticulated skull elements would be more transportable than intact

Table 1. Classification of 1395 identified bones from the Royal Tyrrell Museum of Palaeontology online records into Voorhies (1969) groups indicating that the majority of the bones are within the most transportable group.

Bone	No. of bones
Group I (most transportable)	
Ribs	272
Unidentified vertebrae	154
Phalanges	125
Caudal vertebrae	117
Dorsal vertebrae	100
Cervical vertebrae	59
Sacral vertebrae	48
Scapulae	34
Ulnae	31
Unguals	22
Sternal plates	13
Pedes	9
Atlas	1
Carpal	1
Total	986
Group II (moderately transportable)	
Metatarsals	61
Femora	56
Humeri	50
Tibiae	40
Limb bones*	27
Ilia	25
Metacarpals	23
Radii	18
Ischia	17
Pubes	10
Astragali	4
Total	331
Group III (least transportable)	
Dentaries	59
Maxillae	10
Jaws*	8
Mandible*	1
Total	78

*Some specimens in the museum records were not identified as a specific element.

skulls and placed within either group I or group II. However, no flume experiment conducted on disarticulated skulls is known; therefore, it was decided that complete removal of skulls and skull elements from this calculation was appropriate.

Bone orientation can indicate the paleostream flow direction (Voorhies 1969; Behrensmeyer 1991; Coard and Dennell 1995), which may suggest the type of environment in which the bones were interred (Eberth et al. 2007). A rose diagram can determine if there is a preferred orientation of the long bones (Fiorillo 1988*a*), and one was constructed from the orientations of 1188 bones from the *Pachyrhinosaurus* bone bed (Fig. 2).

One confusing aspect of this site is that large skulls were found lying on top of and next to densely packed smaller elements. Large heavy skulls such as these would require a flow velocity of >150 cm/s to move them downstream (Behrensmeyer 1975). Such a high velocity flow should easily entrain smaller elements and remove them from the site (Behrensmeyer 1991). The smaller elements may have been trapped in the quieter waters downstream of the large elements, or between them. However, there is no consistent pattern in the bone bed map to suggest that this happened. Observations during preparation on some skulls noted small bones jammed within the cavities, and that deep within these recesses, the matrix changed to a clean light grey mudstone from the usual dark grey carbonaceous silty mudstone with clay balls and amber.

Conclusions

The bone orientation and the sediments associated with the bone bed suggest there was a fluvial influence in the formation of the Pipestone Creek bone bed. To determine if it is an allochthonous or autochthonous assemblage, the identified bones (excluding skulls and skull elements) were classified into Voorhies (1969) groups. Of the 1395 elements utilized within this exercise, 70.7% of the elements fall within group I (most transportable), 23.7% are in group II, and only 5.6% are in group III (least transportable) (Table 1). This suggests that the bones were transported from the site of original deposition. This is supported by the completely disarticulated state of the elements, with the exception of some fused adult skulls and skull elements.

Orientations were calculated for 1188 bones identified as complete on the composite quarry map. The resulting rose diagram (Fig. 2) indicates there is no single preferred orientation to the bones and therefore no single directionality to the paleostream flow. However, there is a strong trend to the east, as well as along the north–south axis. This may indicate turbulent flow that would have desposited the bones in many positions (Fiorillo 1988*b*) along with the abundant wood and amber that is found in the site (Currie et al. 2008). The high density of bone present may have inhibited the current's ability to orient bone preferentially. In many cases the bones are in direct contact with each other, allowing any one bone to influence or affect the ability of the surrounding bones to become aligned with the current direction. Other ceratopsian bone bed have shown preferred orientations (Eberth and Getty 2005). Unlike the Pipestone Creek bone bed, the density of bone was lower at these sites and therefore a meaningful comparison of orientation data cannot be completed. The site map is too dense with bone outlines to allow for proper identification of most elements. This excluded the sorting of bones by type, which may have suggested whether certain elements had become preferentially entrained and oriented.

The presence of a different matrix deep within the recesses of the skulls may indicate that the skeletons were previously defleshed and buried at a different location further upstream. The distances involved between first internment and reworking and transport to the Pipestone Creek bone bed site could not have been very far as indicated by the numerous smaller and juvenile elements that would have been winnowed from the site had there been strong flow velocities.

Acknowledgments: The first author thanks Dr. Jim Gardner and Jackie Wilke of the Collections Management section of the Royal Tyrrell Museum for access to the original maps and collections records and Corinne Pugh in the Museum library for help with photocopying. Thanks also to Dr. Philip J. Currie (University of Alberta) for the opportunity to publish these data.

References

Behrensmeyer, A.K. 1975. The taphonomy and paleoecology of Plio-Pleistocene vertebrate assemblages east of Lake Rudolf, Kenya. Bulletin of the Museum of Comparative Zoology, **146**: 473–578.

Behrensmeyer, A.K. 1991. Terrestrial vertebrate accumulations. *In* Taphonomy: releasing the data locked in the fossil record. *Edited by* P.A. Allison and D.E.G. Briggs. Plenum Press, New York. pp. 291–335.

Coard, R. and Dennell, R.W. 1995. Taphonomy of some articulated skeletal remains: transport potential in an artificial environment. Journal of Archaeological Science, **22**: 441–448.

Currie, P.J., Langston, W., Jr., and Tanke, D.H. 2008. A new species of *Pachyrhinosaurus (*Dinosauria, Ceratopsidae) from the Upper Cretaceous of Alberta, Canada. *In* A new horned dinosaur from an Upper Cretaceous bone bed in Alberta. NRC Research Press, Ottawa, Ont., Canada. pp. 1–108.

Eberth, D.A. and Getty, M.A. 2005. Ceratopsian bonebeds: occurrence, origins, and significance. In Dinosaur Provincial Park: a spectacular ancient ecosystem revealed. Edited by P.J. Currie and E.B. Koppelhus. Indiana University Press, Bloomington, Indiana. pp. 501-536.

Eberth, D.A., Rogers, R.R., and Fiorillo, A.R. 2007. A practical approach to the study of bonebeds. *In* Bonebeds: genesis, analysis and paleobiological significance. *Edited by* R.R. Rogers, D.A. Eberth, and A.R. Fiorillo. University of Chicago Press, Chicago, Ill. pp. 265–331.

Fiorillo, A.R. 1988*a*. A proposal for graphic presentation of orientation data from fossils. University of Wyoming, Contributions to Geology, **26**: 1–4.
Fiorillo, A.R. 1988*b*. Taphonomy of Hazard Homestead Quarry (Ogallala Group), Hitchcock County, Nebraska. University of Wyoming, Contributions to Geology, **26**: 57–97.
Tanke, D. 2004. Mosquitoes and mud: the 2003 Royal Tyrrell Museum expedition to the Grande Prairie region (northwestern Alberta, Canada). Alberta Palaeontological Society Bulletin, **19**: 3–31.
Tanke, D. 2005. The Late Cretaceous pachyrhinosaur bonebed (Late Cretaceous: Wapiti Formation) near Grande Prairie, Alberta. *In* Alberta Palaeontological Society, 9th Annual Symposium, Abstracts Volume. Alberta Palaeontological Society, Calgary, Alta. pp. 33–35.
Tanke, D. 2006. Sixty years of pachyrhinosaur (Dinosauria: Ceratopsidae) discoveries in North America. *In* Alberta Palaeontological Society, 10th Annual Symposium, Abstracts Volume. Alberta Palaeontological Society, Calgary, Alta. pp. 38–56.
Voorhies, M.R. 1969. Taphonomy and population dynamics of an Early Pliocene vertebrate fauna, Knox County, Nebraska. University of Wyoming, Contributions to Geology Special Paper, 1.

3. Structure of the brain cavity and inner ear of the centrosaurine ceratopsid dinosaur *Pachyrhinosaurus* based on CT scanning and 3D visualization

LAWRENCE M. WITMER AND
RYAN C. RIDGELY

Abstract

Information on the structure of the brain, cranial nerves, encephalic vasculature, and inner ear of the centrosaurine ceratopsid dinosaur *Pachyrhinosaurus lakustai* (Upper Cretaceous, Wapiti Formation, Alberta, Canada) is presented based on computed tomographic scanning of an isolated braincase followed by 3D visualization. The resulting digital cranial endocast and endosseous labyrinth are the most complete for any ceratopsid, to date, and are compared with physical and digital endocasts of other ceratopsians. In general, the structure of the brain of *P. lakustai*, as inferred from the endocast, was relatively primitive and much more like extant nonavian diapsid endocasts than like endocasts of archosaur groups with derived brains, such as pterosaurs, hadrosaurs, or coelurosaurs (including extant birds). Total brain size is relatively small, and none of the externally discernable brain regions (e.g., olfactory bulbs, cerebrum, cerebellum, optic lobes) are expanded. Currently, there are insufficient data to determine whether the simple structure and small size of the brain of *P. lakustai* are truly plesiomorphic or represent apomorphic reduction (evolutionary

Note: An animation of the 3D visualization of the digital endocast is available on NRC Research Press' Web site at http://pubs.nrc-cnrc.gc.ca/eng/books/books/9780660198194.html.

reversal). The endosseous labyrinth reveals a short cochlea, suggesting that airborne sounds were not particularly important behaviorally. The semicircular canals of the labyrinth are actually somewhat more elongate compared with what little is known of other neoceratopsian labyrinths. The fossil specimen is slightly compressed transversely, and this plastic deformation was subsequently ameliorated by a simple digital retrodeformation that provides a realistic view of brain size and shape. The most significant biological result of the project is that, even accounting for any deformation, the sensorineural and cognitive capabilities of *Pachyrhinosaurus* were modest, certainly in comparison with some other dinosaur clades, such as hadrosaurids and coelurosaurian theropods.

Key words: Dinosauria, Ceratopsia, *Pachyrhinosaurus*, brain, behavior, endocast, CT scanning.

Résumé

Cette étude présente des données sur la structure du cerveau, des nerfs crâniens, du système vasculaire crânien et de l'oreille interne du dinosaure *Pachyrhinosaurus lakustai* (centrosaure cératopsien – Crétacé supérieur, Wapiti Formation, Alberta, Canada), obtenues par tomodensitométrie d'une boîte crânienne. Une analyse par visualisation en trois dimensions est également offerte. Les contre-empreintes (moulage) numériques du crâne et du labyrinthe endo-osseux de ce cératopsien, qui sont les plus détaillées à ce jour, sont comparées aux données physiques et numériques d'autres cératopsiens. En général, la structure du cerveau de *P. lakustai* est relativement primitive et s'apparente davantage aux contre-empreintes des reptiles diapsides existants plutôt qu'à celles du cerveau dérivé d'archosauriens (ptérosaures, hadrosaures, coelurosaures, y compris les oiseaux actuels). Le cerveau est relativement petit et aucune de ses régions externes (bulbes olfactifs, cerveau, cervelet, lobes optiques) n'est élargie. Par ailleurs, les données sont actuellement insuffisantes pour établir si la structure simplifiée et la petite taille du cerveau de *P. lakustai* sont véritablement plésiomorphiques ou représentent plutôt une réduction apomorphique (renversement évolutif). L'examen du labyrinthe endo-osseux révèle une cochlée réduite, ce qui suggère que les sons aériens jouaient un rôle accessoire sur le plan comportemental. Les canaux semi-circulaires du labyrinthe sont en fait relativement plus allongés par rapport aux autres labyrinthes néocératopsiens, bien qu'encore peu connus. Le spécimen fossile est légèrement comprimé dans l'axe transversal, et cette déformation plastique a été subséquemment améliorée par une simple rétrodéformation numérique offrant ainsi une image réaliste de la taille et de la forme du cerveau. Sur le plan biologique, le résultat le plus important du projet est que, même en tenant compte de toute déformation éventuelle, les capacités cognitives et neurosensorielles de *Pachyrhinosaurus* étaient rudimentaires, surtout par rapport à d'autres clades de dinosaures comme les hadrosauridés et les thérapodes coelurosauriens.

Mots-clés : Dinosauria, Ceratopsia, *Pachyrhinosaurus*, cerveau, comportement, contre-empreinte, moulage, tomographie par ordinateur.

[Traduit par la Rédaction]

Introduction

Computed tomographic (CT) scanning has revolutionized the study of brain evolution in extinct archosaurs, such as dinosaurs, its X rays digitally slicing through bone and stone to allow access to the endocranial cavity. CT has provided new insights into the brain structure of pterosaurs (Witmer et al. 2003), ankylosaurians (Witmer and Ridgely 2008), sauropods (Sereno et al. 2007; Witmer et al. 2008), and theropods (Rogers 1998, 1999, 2005; Brochu 2000, 2003; Larsson 2001; Franzosa 2004; Franzosa and Rowe 2005; Sampson and Witmer 2007; Kundrát 2007), including birds (Domínguez Alonso et al. 2004; Kurochkin et al. 2006, 2007). With the exception of a CT-based study of the basal ceratopsian *Psittacosaurus* (Zhou et al. 2007), the study of the brain structure of horned dinosaurs has not made the jump into the digital age, probably in part because the braincase tends to be attached to skulls that are simply too large to fit through most CT scanners. Here we report on the first CT-based study of ceratopsid endocranial structure, producing the most complete view to date for any ceratopsian. The disarticulated braincase of the centrosaurine *Pachyrhinosaurus lakustai*, derived from the Pipestone Creek bone bed (Grande Prairie, Alberta; see Currie et al. 2008), forms the basis of this study.

Some previous studies of ceratopsid endocranial structure have drawn on observations of broken specimens, whereas others have made latex or plaster casts of the endocranial surface from specimens from which the rock matrix had been removed (known as brain casts or cranial endocasts). The endocranial morphology of only a few neoceratopsian taxa has been reported in the literature (see Hopson (1979) for an excellent review): *Protoceratops* (Brown and Schlaikjer 1940); the chasmosaurine ceratopsids *Anchiceratops* (Brown 1914) and *Triceratops* (Burckhardt 1892; Marsh 1896; Hay 1909; Gilmore 1919; Forster 1996); and of significance for this paper, the centrosaurine ceratopsid *Pachyrhinosaurus canadensis* (Langston 1975). Of these reports, only Marsh (1896), Brown (1914), Brown and Schlaikjer (1940), and Forster (1996) provided data based on cranial endocasts. Thus, the findings presented here will be not only the first CT-based study but also the first presentation of a centrosaurine endocast and inner ear (Figs. 1–3).

Beyond simply affording "a look inside," CT scanning provides a number of advantages for studying the endocranial region. For example, because CT scanning transfers the specimen into the digital realm, it provides new opportunities for visualization and analysis. Structures such as the brain endocast, cranial nerve trunks, vascular canals, and endosseous labyrinth can be digitally extracted (a process often termed "segmentation") and viewed in isolation or even in situ within the surrounding bone, which can be rendered transparent. This latter attribute can be useful for clarifying the identity of various apertures on the external surface of the braincase in that, whereas the external surface can be highly variable in form (due to the variable functional constraints to which this surface must respond), the internal surface of the braincase is evolutionarily conservative between even distantly related taxa. The ability to unambiguously trace canals through the bony walls of the braincase permits us to correct some previous identifications made on the basis of only external morphology. The jump into the digital realm also makes quantification a trivial matter, allowing the rapid and accurate determination of parameters such as endocast volume, etc. Moreover, the data can be digitally manipulated to ameliorate the effects of postmortem distortion. For example, Currie et al. (2008) reported that the *P. lakustai* braincase at hand is somewhat laterally compressed; we used a simple digital transformation to address the distortion and arrive at a more realistic representation of the braincase and endocast, as well as perhaps more accurate metrics.

Institutional abbreviations: AMNH, American Museum of Natural History, New York, USA; IGM, Mongolian Institute of Geology, Ulaan Bataar, Mongolia; TMP, Royal Tyrrell Museum of Palaeontology, Drumheller, Alberta, Canada; USNM, United States National Museum of Natural History, The Smithsonian Institution, Washington, DC.

Materials and Methods

A single isolated braincase of *Pachyrhinosaurus lakustai* (TMP 1989.55.1243) was available for CT scanning. This specimen derives from the Pipestone Creek bone bed in west–central Alberta, in Upper Cretaceous (Upper Campanian to Maastrichtian) rocks of the Wapiti Formation (Currie et al. 2008). Based on comparison with braincases of other specimens from the bone bed, Currie et al. (2008) regarded this somewhat compressed transversely specimen as coming from a relatively small individual. Indeed, the degree of fusion of the sutures between bones suggests that the individual was mature but relatively young. Although the results will not be presented here, we have drawn on comparative analysis of CT scan data, digital endocasts, and endosseous labyrinths generated as part of a larger project. Relevant ceratopsian specimens in this larger sample include a *Psittacosaurus* skull (IGM 100/1132), an unnumbered IGM "protoceratopsian" skull, and an unnumbered TMP braincase of *Centrosaurus*. Moreover, the AMNH collection of physical endocasts was studied, including the *Protoceratops* endocast (AMNH 6466) published by Brown and Schlaikjer (1940) and Hopson (1979), the *Anchiceratops* endocast (AMNH 5259) published by Brown (1914) and Hopson (1979), and the *Triceratops* endocast (USNM 4286) published by Hay (1909).

TMP 1989.55.1243 was scanned on 7 April 2005 at Canada Diagnostic Centre in Calgary, Alberta, on a General Electric LightSpeed Ultra Multislice CT scanner. It was scanned helically at a slice thickness of 1.3 mm, 140 kV, and 300 mA, and was reconstructed using both standard and bone algorithms. Data were output from the scanner in DICOM format and then imported into Amira 4.1.2 (Mercury-TGS, Chelmsford, Massachusetts) for viewing, analysis, and visualization on 32- and 64-bit PC workstations with 4 GB of RAM and nVidia Quadro FX 3000 or 4500 video cards and running Microsoft Windows XP Professional and Windows XP Professional x64. Anatomical features of interest (e.g., cranial endocast, labyrinth, blood vessels, etc.) were highlighted and digitally extracted using Amira's segmentation tools for quantification and visualization. Both surfaces and volumes were generated and were used to illustrate this paper. To facilitate discussion, we will refer to the digital casts of structures as if they were the structures themselves (e.g., "carotid canal" versus "digital cast of carotid canal"). Thus, bony structures and the equivalent structures on the cranial endocast will be referred to by the same name, despite topologically having the relationship of glove to hand. Movies and interactive animations of the digital endocast and braincase are available on the authors' website (www.ohio.edu/witmerlab).

Although quantitative metrics of various structures (e.g., volumes, areas, lengths) are easily generated by the software, we choose not to report those numbers here (with the exception of endocast volume, which is reported in the Discussion in the heuristic context of retrodeformation). Instead, we use relative, more qualitative, characterizations of dimensions. The reason for this decision stems mostly from the fact that the braincase at hand is a disarticulated specimen for which a body mass or other general size metric is unavailable (setting aside the problems associated with reliably estimating body mass in extinct animals). Also, the specimen is sufficiently distorted that the precise values may not be trustworthy, although the general proportions likely are. Finally, different researchers use different measurement schemes, and thus, we hesitate to introduce values into a literature already cluttered with metrics of doubtful comparability. That being said, all metric data are available on request from the authors, as are the digital models from which new measurements can be made.

Note: An animation of the 3D visualization of the digital endocast is available on NRC Research Press' Web site at http://pubs.nrc-cnrc.gc.ca/eng/books/books/9780660198194.html.

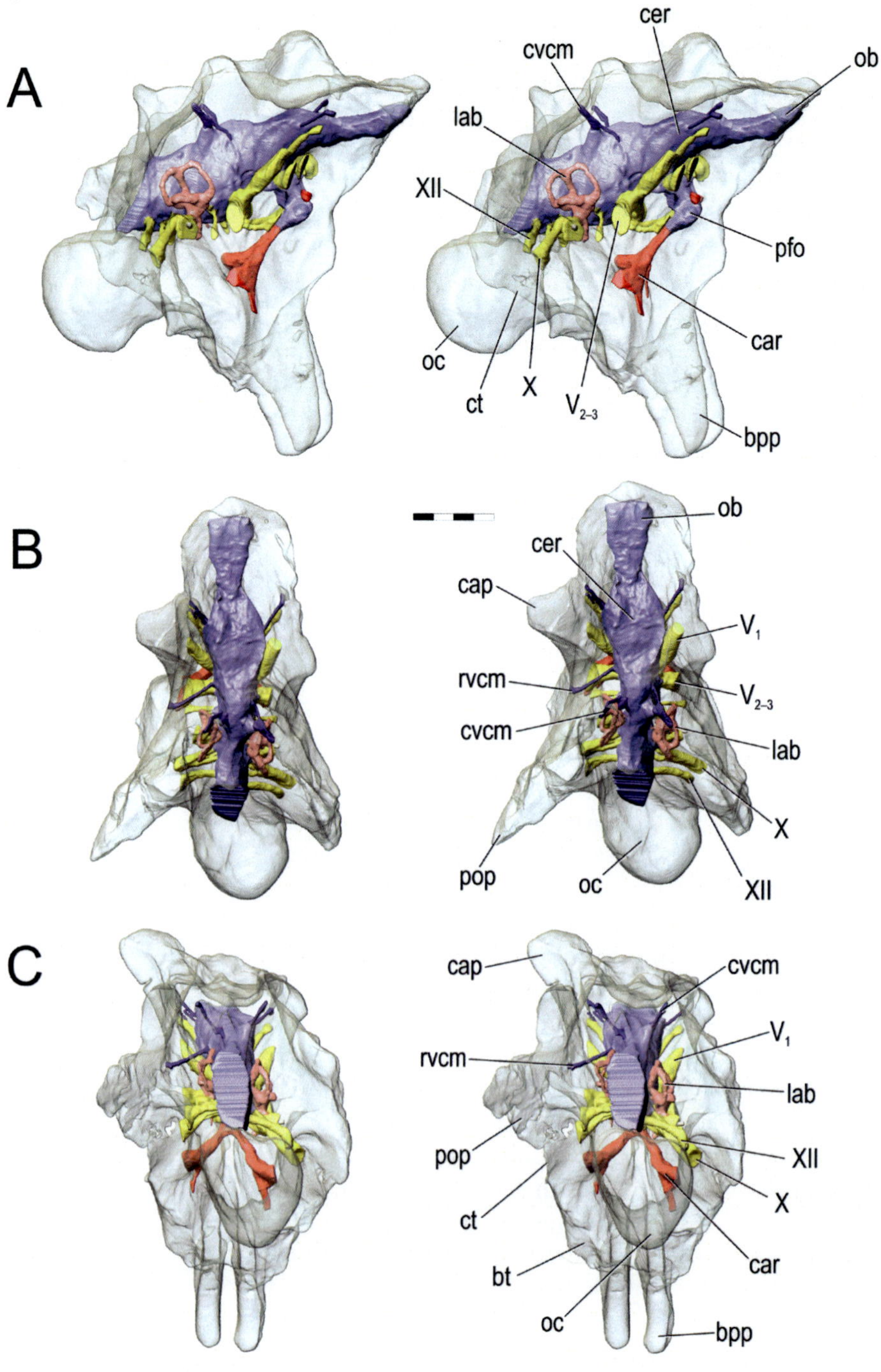

Fig. 1. Stereopairs of articulated braincase of Pachyrhinosaurus lakustai *(TMP 1989.55.1243) derived from reconstructed computed tomographic (CT) scans and shown in the following views: (A) right lateral, (B) dorsal, (C) caudal, (D) left lateral, (E) ventral, and (F) rostral. Bone is rendered semitransparent, revealing cranial endocast in blue, endosseous labyrinth in pink, nerve canals (most of which also transmit veins) in yellow, smaller venous canals in dark blue, and arterial canals in red. Scale bar = 4 cm. Abbreviations: bpp, basipterygoid process; bt, basal tuber; cap, capitate process of laterosphenoid; car, cerebral carotid artery canal; cer, cerebral hemisphere; ct, crista tuberalis; cvcm, caudal middle cerebral vein; lab, endosseous labyrinth; ob, olfactory bulb; oc, occipital condyle; pfo, pituitary (= hypophyseal) fossa; pop, paroccipital process; rvcm, rostral middle cerebral vein; V_1, ophthalmic nerve canal; V_{2-3}, maxillomandibular nerve canal; X, shared canal for vagus and accessory nerves and accompanying vessels; XII, hypoglossal nerve canal.*

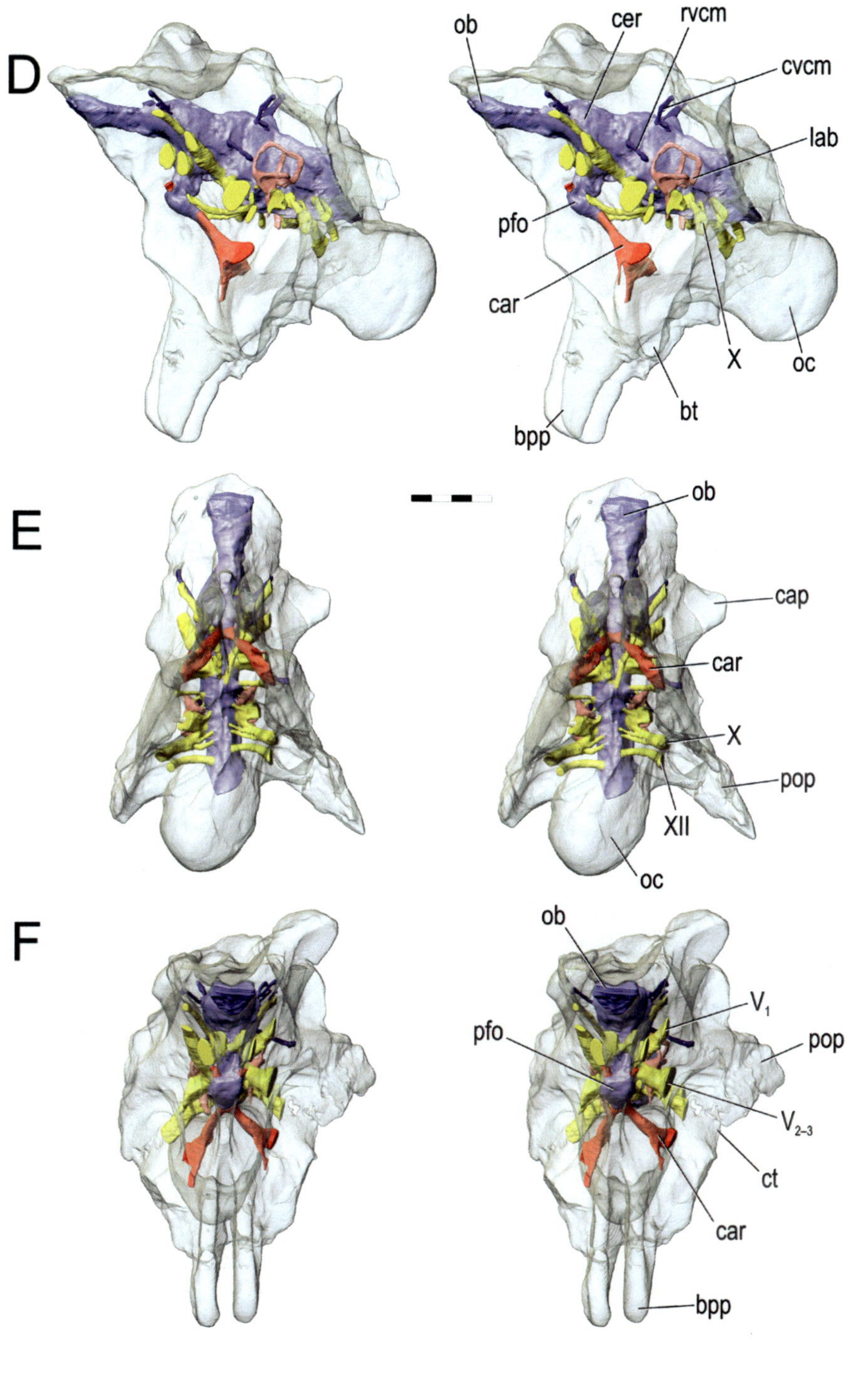
D
ob
cer
rvcm
cvcm
lab
pfo
car
X
oc
bt
bpp
E
ob
cap
car
X
pop
XII
oc
F
ob
V_1
pfo
pop
V_{2-3}
ct
car
bpp

Figs. 2A–E

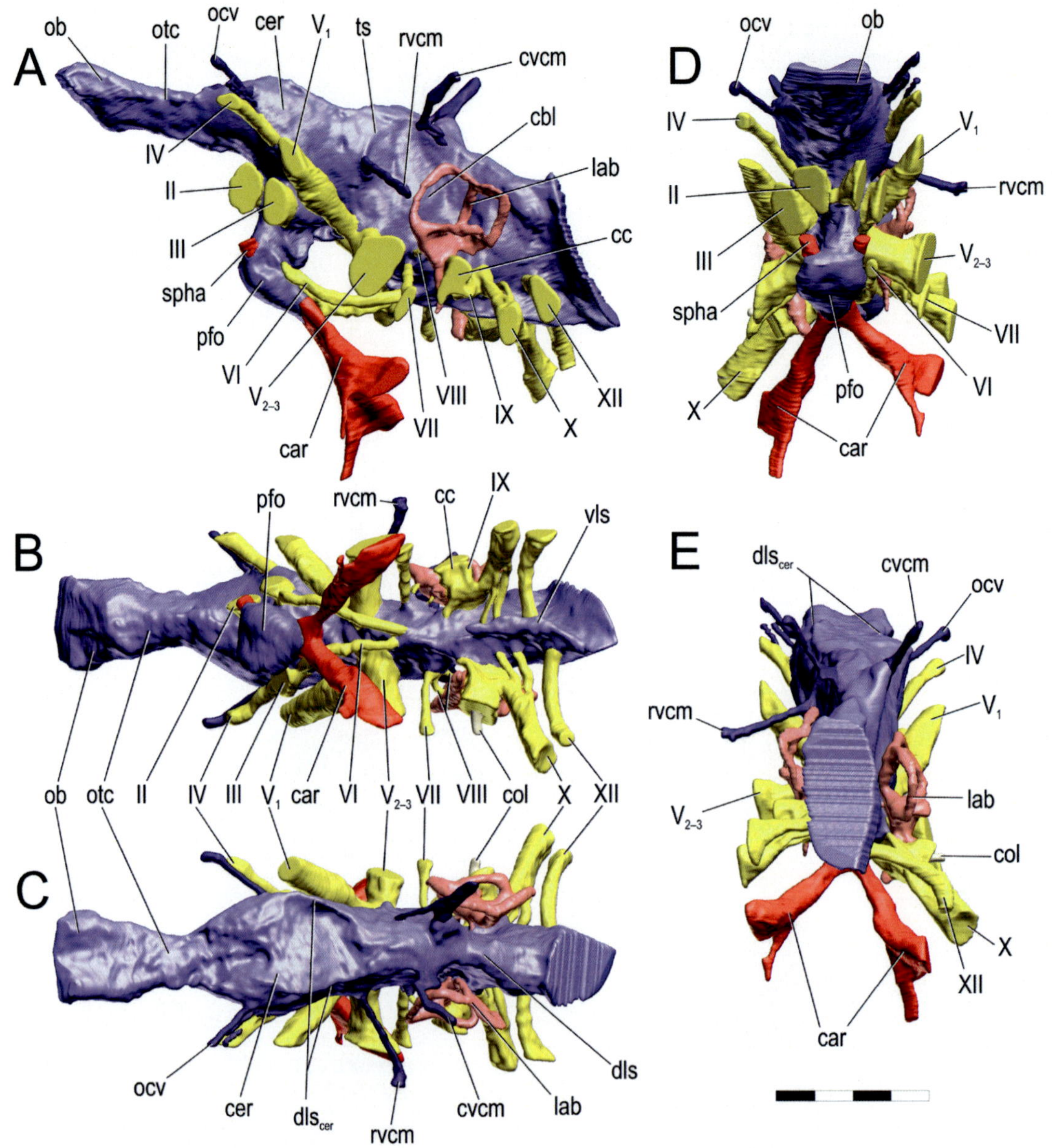

Fig. 2. Cranial endocast of Pachyrhinosaurus lakustai *(TMP 1989.55.1243) reconstructed from computed tomographic (CT) scans and shown in the following views: (A) left lateral, (B) ventral, (C) dorsal, (D) rostral, and (E) caudal. Stereopairs are in the following views: (F) left lateral, (G) right lateral, (H) dorsal, (I) ventral, (J) caudal, and (K) rostral. Color scheme: cranial endocast, blue; endosseous labyrinth, pink; nerve canals (most of which also transmit veins), yellow; smaller venous canals, dark blue; arterial canals, red. Scale bar = 4 cm. Abbreviations: car, cerebral carotid artery canal; cbl, cerebellum; cc, columellar canal; cer, cerebral hemisphere; cvcm, caudal middle cerebral vein; col, columella (= stapes); dls, dorsal longitudinal sinus (a dural venous sinus); dls_{cer}, cerebral branch of dorsal longitudinal sinus; lab, endosseous labyrinth; ob, olfactory bulb; ocv, orbitocerebral vein canal; otc, olfactory tract cavity; pfo, pituitary (= hypophyseal) fossa; rvcm, rostral middle cerebral vein; spha, sphenoid artery canal; ts, transverse sinus; vls, ventral longitudinal sinus (a dural venous sinus); II, optic nerve canal; III, oculomotor nerve canal; IV, trochlear nerve canal; V_1, ophthalmic nerve canal; V_{2-3}, maxillomandibular nerve canal; VI, abducens nerve canal; VII, facial nerve canal; VIII, vestibulocochlear nerve canal; IX, glossopharyngeal nerve canal (part of columellar canal); X, shared canal for vagus and accessory nerves and accompanying vessels; XII, hypoglossal nerve canal.*

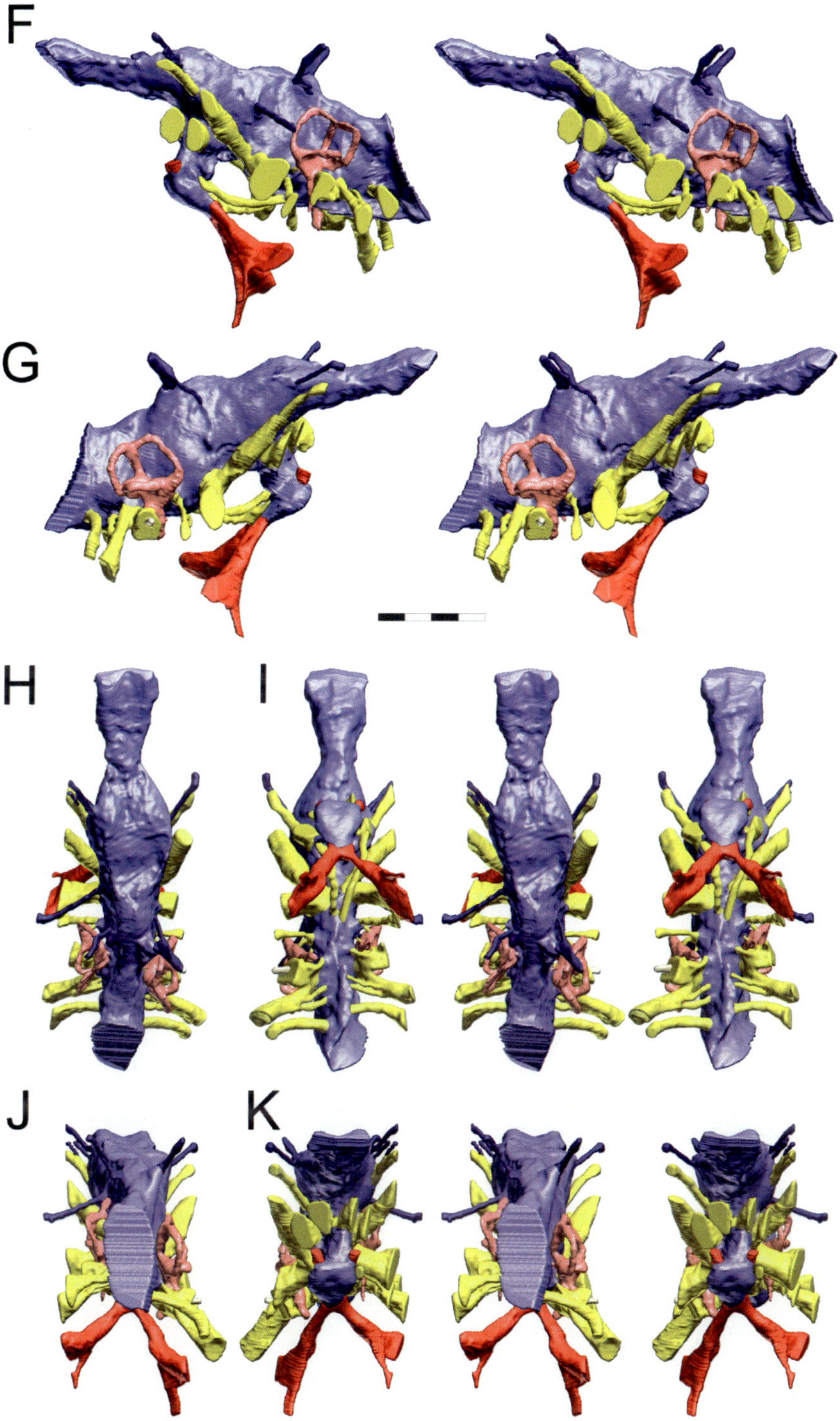
F
G
H
I
J
K

Fig. 3. Endosseous labyrinth of the left inner ear of Pachyrhinosaurus lakustai *(TMP 1989.55.1243) reconstructed from computed tomographic (CT) scans. Stereopairs are in the following views: (A) lateral, (B) caudal, and (C) dorsal. Orientations were determined based on orientation of the labyrinth within the skull and with the lateral semicircular canal placed horizontally. Scale bar = 4 cm. Abbreviations: c, cochlea (= lagena); crc, crus communis; csc, caudal (posterior vertical) semicircular canal; ed, endolymphatic duct; fp, fenestra perilymphaticum (= round window); fv, fenestra vestibuli (= oval window); lsc, lateral (horizontal) semicircular canal; lsca, ampulla of lateral semicircular canal; rsc, rostral (anterior vertical) semicircular canal; rsca, ampulla of rostral semicircular canal; ve, vestibule of inner ear.*

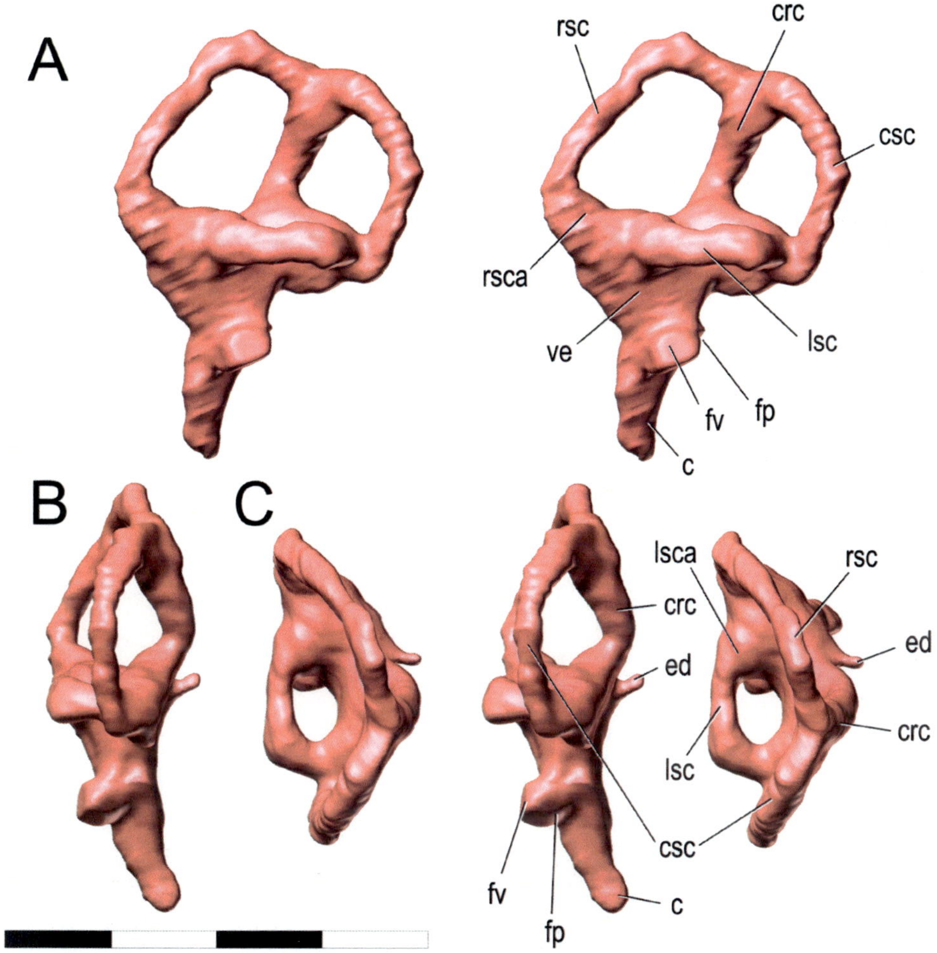
A
rsc
crc
csc
rsca
ve
lsc
fv
fp
c
B
C
lsca
crc
rsc
ed
ed
crc
lsc
csc
fv
fp
c

Fig. 4A

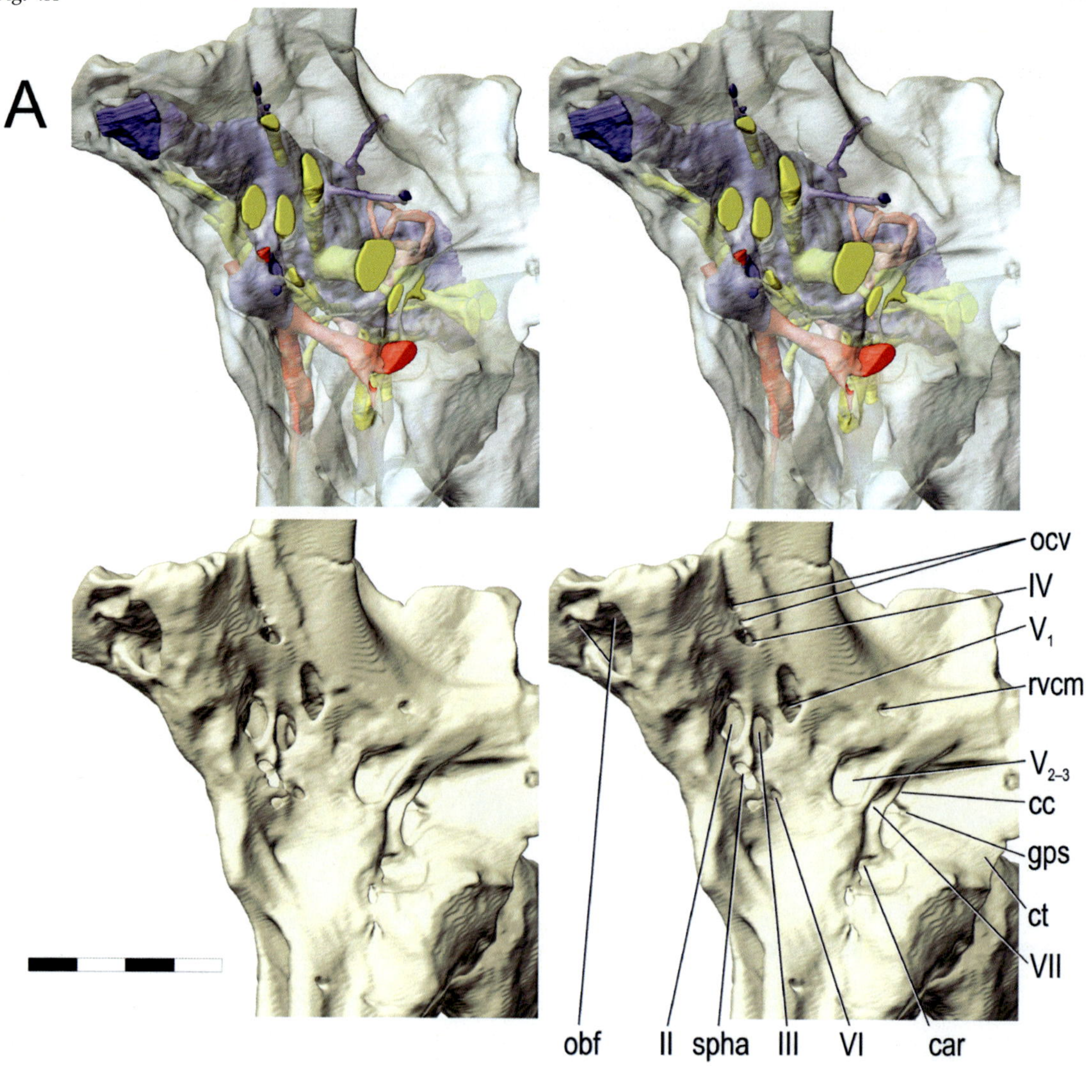

Fig. 4. Stereopairs of a braincase of Pachyrhinosaurus lakustai *(TMP 1989.55.1243) reconstructed from computed tomographic (CT) scans in the (A) left rostroventrolateral view and (B) left caudolateral view to show the osteological correlates (e.g., foramina, fossae, crests) of many of the soft-tissue structures discussed in the text. Each view consists of a set of stereopairs (above) showing a semitransparent braincase revealing enclosed soft-tissue structures, coupled with a set of stereopairs (below) in the same view showing the bony braincase and labeled structures. Color scheme: cranial endocast, blue; endosseous labyrinth, pink; nerve canals (most of which also transmit veins), yellow; smaller venous canals, dark blue; arterial canals, red. Scale bar = 4 cm. Abbreviations: bt, basal tuber; car, cerebral carotid artery canal; cc, columellar canal; ct, crista tuberalis; cvcm, caudal middle cerebral vein; fm, foramen magnum; gps, glossopharyngeal sulcus; obf, olfactory bulb fossa; ocv, orbitocerebral vein canal; rvcm, rostral middle cerebral vein; spha, sphenoid artery canal; II, optic nerve canal; III, oculomotor nerve canal; IV, trochlear nerve canal; V_1, ophthalmic nerve canal; V_{2-3}, maxillomandibular nerve canal; VI, abducens nerve canal; VII, facial nerve canal; X, shared canal for vagus and accessory nerves and accompanying vessels; XII, hypoglossal canal.*

Fig. 4B

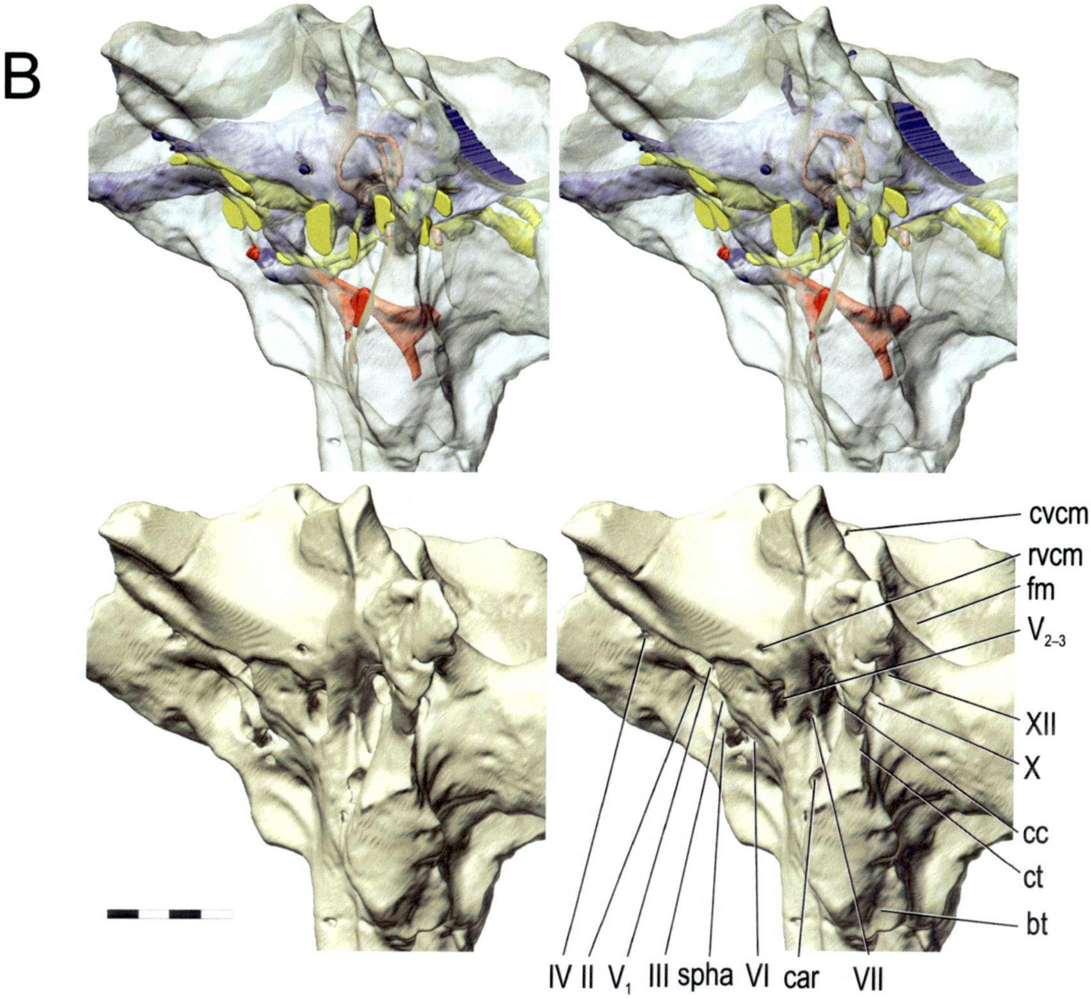

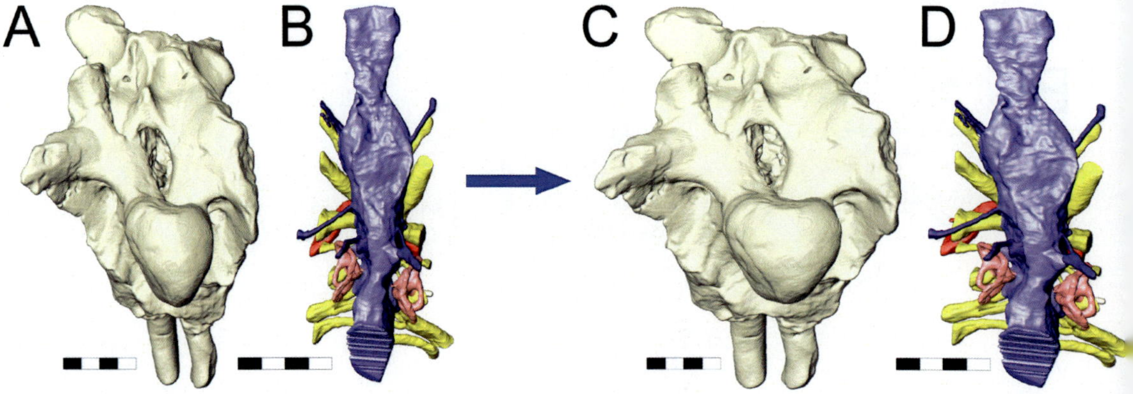

Fig. 5. Simple retrodeformation of a braincase and cranial endocast of Pachyrhinosaurus lakustai *(TMP 1989.55.1243) reconstructed from computed tomographic (CT) scans. (A, C) Braincase in caudal view. (B, D) Cranial endocast in dorsal view. Scale bar = 4 cm. In A and B, the structures are shown as preserved prior to retrodeformation. In C and D, the structures have been transformed by simply stretching the dataset transversely so that the width of the occipital condyle equals its height (the condition observed in undeformed specimens). Even this simple retrodeformation restores a more natural shape to the braincase and endocast and increases endocast volume by 35%.*

Results

The CT scan data for TMP 1989.55.1243 were excellent, allowing clear discrimination of fossil bone, rock matrix, and the plaster used to reconstruct the missing portions (e.g., left basipterygoid process). Currie et al. (2008) provided a complete osteological description and illustration of this specimen, and thus this paper will focus on the cranial endocast, endosseous labyrinth, and endocranial vasculature. Nevertheless, some aspects of the bony braincase will be discussed, as they pertain to the reconstructed soft tissues, and the position of these soft-tissue systems are illustrated in place within the semitransparent braincase in Figure 1. Rather than provide a lengthy description of the structures, we will allow the illustrations to convey information on general form and relationships and instead focus on comparisons and significance.

Cranial endocast

The term "cranial endocast" (or just "endocast") is generally used for a replica (whether virtual or physical) of the internal surface of the brain cavity. As discussed by many other workers (e.g., Jerison 1973; Hopson 1979), an endocast does not always faithfully record the shape of the underlying brain because the brain may not sufficiently fill the endocranium. In that case, the endocast constitutes a replica of the envelope of the dura mater (Osborn 1912). In some archosaurs, such as pterosaurs (Witmer et al. 2003), derived coelurosaurs (Currie 1985, 1995; Currie and Zhao 1993; Osmólska 2004; Kundrát 2007), and perhaps some ornithischians (Evans 2005), the brain resembled that of extant birds in that it largely filled the endocranium to the extent that an endocast fairly represents the general brain structure (Iwaniuk and Nelson 2002). However, the endocast of *P. lakustai* (Figs. 1, 2) is more like that of extant reptiles (and indeed most fossil archosaurs; Hopson 1979; Witmer et al. 2008) in that the endocast is not particularly brain-like in form, suggesting that the brain itself was perhaps markedly smaller than the endocast. Again, the endocast is somewhat distorted due to plastic deformation of the braincase (see above), but the transverse compression has not grossly altered its morphological features.

As is typically the case in archosaurs, the telencephalic portions of the brain are the most interpretable, presumably due to there having been less intervening tissue between the brain and bone. Thus, the cerebral hemispheres are clearly visible as rounded swellings in the forebrain region. Likewise, the paired olfactory bulbs are observable rostrally as they diverge from the median cavity for the olfactory tracts (Fig. 2C). Viewed in place (Fig. 1), the olfactory tract cavity can be seen to run through the dorsomedial roof of the orbit such that the olfactory bulbs are situated at the back of the nasal cavity. The pituitary region is also clearly visible (Figs. 2A, 2B), but the cast of the bony pituitary fossa does not provide a reliable measure of the size and shape of the hypophysis itself because the fossa houses a number of other structures as well, such as the cerebral carotid artery and cavernous venous sinus. Nevertheless, the pituitary bears a well-marked infundibular constriction below the base of the oculomotor nerve trunk, and the structure generally resembles that illustrated for the endocasts of *Triceratops* (Marsh 1896; Hay 1909; Forster 1996) and *Anchiceratops* (Brown 1914; Hopson 1979). The cerebellum also is represented in the endocast as a low swelling (Fig. 2A). In extant diapsids, the cerebellum is typically located caudal to the transverse sinus system (including the rostral and caudal middle cerebral veins), between the paired rostral semicircular canals, and dorsal to the proximal trunk of the trigeminal nerve (Sampson and Witmer 2007; Ridgely and Witmer 2007). The swelling on the endocast of *P. lakustai* is in precisely this location, and a similar cerebellar swelling is present in the *Centrosaurus* endocast. Based on the size of the swelling, and even accounting for postmortem deformation, the cerebellum must have been relatively small in size. There is no indication in the endocast that the cerebellum had a floccular lobe (cerebellar auricle).

Unlike the brain components just noted, the position of the optic tectum (lobe) is not clearly visible. Hay (1909) and Forster (1996) identified low swellings on their endocasts of *Triceratops* as pertaining to the optic tectum, although Hopson (1979) regarded the tecta as not being visible in his ceratopsid endocasts, and the latter is likewise true for *P. lakustai.* Based on comparisons with extant diapsids (Ridgely and Witmer 2007), the optic tectum should be located in the region just caudal to the cerebral swelling and just rostral to the transverse sinus system, but there is no discernable swelling in this location in the endocast of TMP 1989.55.1243. This situation suggests that the optic tectum was relatively small, although the influence of postmortem deformation cannot be completely excluded. It moreover suggests that brain organization in *P. lakustai* resembled extant crocodylians and other reptiles in having an "in-line" brain with the cerebral hemispheres, optic tecta, and cerebellum in a rostrocaudal sequence and the contralateral optic tecta contacting each other. On the other hand, in more encephalized archosaurs, such as pterosaurs and many advanced theropods (especially birds), the optic tecta were displaced ventrolaterally and did not contact each other dorsally (Currie and Zhao 1993; Witmer et al. 2003; Kundrát 2007).

The cranial endocast preserves a number of vascular elements, and although these vessel traces obscure some of the underlying neural components, they also provide evidence for the gross limits of some of the brain parts. For example, the transverse sinus system was discussed in the context of providing a rough boundary between the optic tectum and cerebellum. The transverse sinus (a dural venous sinus) is visible on the endocast as a low curving ridge (Fig. 2A). It is continued laterally and caudodorsally as the rostral and caudal middle cerebral veins, respectively, the lateral branch opening in the skull just dorsal to the maxillomandibular foramen and the caudal branch opening on the occiput (Figs. 1, 4). Another dural sinus that is clearly visible is the ventral longitudinal sinus. Identified earlier in the theropod *Majungasaurus* (Sampson and Witmer 2007), the ventral longitudinal sinus is a common feature of archosaur endocasts, forming a median ridge, which is long in *P. lakustai*, extending the length of the brainstem (Fig. 2B). The upper "shoulders" of the ventral longitudinal sinus represent the ventral margin of the medulla. The occipital dural venous sinus obscures the dorsal margin of the medulla and grades rostrally into the dorsal longitudinal sinus (Figs. 2A, 2C).

The dorsal longitudinal sinus extends rostrodorsally to the region directly above the cerebellum where it joins with the transverse sinus system. At this point, it splits and diverges, passing rostrally as a ridge along the dorsolateral edge of each cerebral hemisphere. These cerebral branches of the dorsal longitudinal sinus may correspond to the paired "dorsal venous sinuses" identified in *Triceratops* by Forster (1996). Interestingly, this branched dorsal longitudinal sinus is not present in the endocast of *Centrosaurus* but does appear to be present in the physical endocast of *Protoceratops* (AMNH 6466) and the digital endocast of *Psittacosaurus* (IGM 100/1132). In *Psittacosaurus*, the endocast ridges above the cerebral hemispheres are clearly coincident with sutures (e.g., frontal-laterosphenoid), perhaps suggesting that these features are architectural rather than vascular. However, dural venous sinuses positioned within bony sutures are common in archosaurs (Sampson and Witmer 2007; Witmer et al. 2008). Moreover, in *Pachyrhinosaurus lakustai* the ridges do not run entirely within sutures, and in both *Pachyrhinosaurus* and *Psittacosaurus* the endocast ridges are continuous with other, clearly vascular, features. In *Pachyrhinosaurus lakustai*, the cerebral branches of the dorsal longitudinal sinus drain into the orbit via the orbitocerebral vein canals (Figs. 2, 4). As in many other dinosaurs (see Witmer et al. 2008), the orbitocerebral vein(s) provided an important anastomotic connection between orbital veins and the endocranium.

The cerebral carotid artery passed through the typical canal in the basisphenoid on its way to the hypophyseal (pituitary) fossa, each side opening separately into the caudoventrolateral corner of the fossa (Currie et al. 2008; Figs. 1, 2, 4). Emerging from the rostrodorsal aspect of the hypophyseal fossa was a branch of the carotid artery that supplied the orbital contents and dorsum of the palate. Although this same vessel has been called the ophthalmic artery in other ceratopsids (Hay 1909; Brown 1914; Hopson 1979; Forster 1996), we regard this as the sphenoidal or sphenopalatine artery (Sampson and Witmer 2007; Fig. 2), a term from avian anatomy that recognizes the fact that in many birds, the true ophthalmic artery ontogenetically regresses to be replaced by the sphenoidal artery.

The pattern of the cranial nerve trunks is illustrated in Fig. 2 and their emergence externally on the braincase is in Fig. 4. In general, the nerve canals that enter the orbit are relatively large, seemingly much larger than necessary to perform their neural functions (e.g., innervating extraocular muscles). In extant archosaurs, the cranial nerve canals and foramina—particularly the rostral ones (CN II–V)—transmit not only nerves but also veins from the orbit and nasal cavity into the endocranial cavity (Sedlmayr 2002). Thus, it seems likely that *P. lakustai* was like some other dinosaurs (e.g., sauropods; Witmer et al. 2008) in not only having the extant condition but also elaborating it further by expanding the venous component. The basic pattern of the nerves (Fig. 2) largely agrees with that presented by previous workers (e.g., Brown 1914; Hopson 1979; Forster 1996), and indeed the pattern is conservative across major clades (Witmer et al. 2008). Brown's (1914) study of *Anchiceratops* is the only comparable study that treated the cranial nerve canals throughout their entire length. One notable difference between *Pachyrhinosaurus* and *Anchiceratops* is the trigeminal nerve. In both taxa (as in ceratopsians generally; Langston 1975), the ophthalmic branch (CN V_1) and the maxillomandibular branch (CN V_{2-3}) took separate courses through the braincase to emerge at separate foramina (Figs. 1, 2, 4). However, whereas in *Anchiceratops* the two branches share a long common trunk coming off the endocast prior to their distal split, in *P. lakustai* the two trunks come off the main endocast almost separately. *Centrosaurus* shares the *P. lakustai* condition. It is not known whether this condition characterizes just centrosaurines, and it is possible that the *Anchiceratops* condition is a preservational artifact. Current work by the authors on the digital endocast of *Triceratops* should shed light on the chasmosaurine condition.

The facial and vestibulocochlear nerve canals (CN VII and CN VIII, respectively) share the typical common origin within the internal acoustic fossa. The facial nerve is unremarkable, but separate vestibular and cochlear branches of CN VIII were traceable on both sides (Fig. 2B). The caudalmost nerve trunks likewise display a relatively typical pattern. The vagus (CN X) and accessory (CN XI) nerve trunks were more or less indistinguishably joined within the vagal canal, and the glossopharyngeal nerve (CN IX) may well have joined them (but see subsequent discussion). As discussed by Sampson and Witmer (2007), numerous terms have been used for the common canal for these nerves (e.g., vagal, jugular, metotic), and the terminology is further complicated by the proximity of the perilymphatic system rostrally such that another set of terms (e.g., fenestra cochleae, fenestra pseudorotunda) is in play. The vagal canal passes caudal to the crista tuberalis (the bony crest that runs from the paroccipital process to the basal tuber; see Sampson and Witmer 2007) to open on the occiput (Figs. 1, 2, 4). The vagal canal receives the two rostralmost nerve canals of the three hypoglossal nerve canals, leaving the caudal-most—and by far the largest—hypoglossal trunk to open externally by way of its own foramen (Figs. 2, 4). Having three hypoglossal canals with the rostral two joining the vagal canal was reported previously for *Pachyrhinosaurus* (and *Chasmosaurus*) by Langston (1975) and for *Triceratops* by Forster (1996), yet the digital endocast of *Centrosaurus* bears only two hypoglossal canals, the rostral one of which enters the vagal canal.

Whereas the vagal canal passes caudal to the crista tuberalis, it is continuous rostrally with a space (a projection on the endocast) that extends rostral to the crista. We have labeled this structure the columellar canal in Figs. 2 and 4 because it clearly houses the columella (= stapes), which is preserved more or less in place on the right side of TMP 1989.55.1243 (Figs. 2B, 2C, 2E). However, it also may have transmitted the glossopharyngeal nerve (CN IX) caudally given that there is a notch in the rostral margin of the crista tuberalis in this area (labeled glossopharyngeal sulcus in Fig. 4A; see also Fig. 2). The course of the glossopharyngeal nerve is variable in archosaurs and depends largely on whether the metotic fissure is divided or not, which it is in Ceratopsia. The structure on the endocast that we call the columellar canal may also contain some venous components but certainly contains portions of the perilymphatic system, which will be mentioned briefly in the next section on the inner ear.

As correctly noted by Langston (1975, p. 1588), the external foramina in the braincase can be difficult to identify because the nerve and vascular canals branch and "sometimes coalesce within the thickened ceratopsian cranial walls." Fortunately, CT scanning and 3D visualization can resolve almost all such uncertainties, and Fig. 4 compares illustrations of the endocast and neurovascular trunks in situ with the isolated braincase so that the foramina can be identified. This approach allows correction of some published accounts that did not have the benefit of CT-based visualization. For example, in the braincase illustration of *Centrosaurus* presented by Dodson et al. (2004, p. 501), the foramen labeled "c.n. VI" should be the ophthalmic branch of the trigeminal nerve (CN V_1). These same authors also illustrate a grouping of three foramina, one of which is labeled "c.n. IV;" actually, the ventralmost opening is the trochlear foramen (CN IV) and the upper two openings are orbitocerebral vein foramina.

Endosseous labyrinth of the inner ear

The labyrinth of the inner ear is well preserved on both sides of TMP 1989.55.1243 (Fig. 3) and provides the best information to date on the inner ear of a ceratopsian (although we now have comparable data for *Psittacosaurus*, *Centrosaurus*, and *Triceratops*, which will be published elsewhere). Physical representations of the inner ears of *Protoceratops* (Brown and Schlaikjer 1940) and *Anchiceratops* (Brown 1914) have been published, having been made along with the cranial endocast. These are informative, but we assume that there must have been some measure of reconstruction involved given that they somehow were physically removed from the fossils. Zhou et al. (2007) presented CT-based data on the inner ear of their specimens of *Psittacosaurus*. Our digital representation of the inner ear (Fig. 3) corresponds to neither the membranous (endolymphatic) labyrinth (which, being soft tissue, is not preserved) nor the osseous (bony) labyrinth but rather corresponds to the inside of the osseous labyrinth. Thus, although it is typically referred to as a digital "osseous labyrinth" (e.g., Sampson and Witmer 2007), we have adopted the more apt term "endosseous labyrinth" (Witmer et al. 2008; see also Sereno et al. 2007).

The endosseous labyrinth of *Pachyrhinosaurus lakustai* (Figs. 2, 3) is of a relatively generalized form. The upper part of the inner ear (the vestibule and semicircular canals) is associated with the sense of balance and equilibrium with important links to eye movements (see Discussion). The semicircular canals of *P. lakustai* are not as thin and elongate as in the labyrinths of *Psittacosaurus* (IGM 100/1132) but are not quite as stocky as those of *Anchiceratops* (AMNH 5259). The rostral semicircular canal is the longest of the three in *Pachyrhinosaurus lakustai* and actually ascends above the level of the common crus. This situation is different from that observed in *Anchiceratops* where the rostral and caudal canals are more similar in length. Likewise, in *Protoceratops* (AMNH 6466) and *Anchiceratops* (AMNH 5259), the rostral canals are lower and descend from the common crus. Still, the vestibular portion of the labyrinth of *Pachyrhinosaurus lakustai* more closely resembles that of these neoceratopsians than it does the gracile, theropod-like ear of *Psittacosaurus*. In particular, the lateral canal of *Pachyrhinosaurus lakustai* is relatively short (Fig. 3C), although not as short as that in sauropods (Witmer et al. 2008).

The lower part of the inner ear, the cochlea, comprises the hearing organ. We regard the cochlea as being that portion distal to the position of the fenestra vestibuli (= fenestra ovalis), which is where the footplate of the columella is seated. The cochlea of *P. lakustai* is relatively short, certainly in comparison to *Psittacosaurus* (IGM 100/1132) in which it is quite long. The form of the cochlea is not clear on the physical endocasts of *Protoceratops* (AMNH 6466) and *Anchiceratops* (AMNH 5259) but probably is comparable to that of *Pachyrhinosaurus lakustai*.

The form of the perilymphatic system is not entirely clear. As noted, the embryonic metotic fissure is divided in ceratopsians, and the portion rostral to the crista tuberalis was partially occupied by the perilymphatic duct. In some archosaurs with divided metotic fissures (e.g., crocodyliforms, advanced theropods), the perilymphatic duct extended caudally where it abutted bony elements to form a secondary tympanic membrane (e.g., fenestra cochleae or fenestra pseudorotunda; see Sampson and Witmer 2007, and references therein). We are not sufficiently certain that these perilymphatic modifications were fully present in *P. lakustai*, and so our labeling in Fig. 3 uses terms more consistent with a primitive state, that is, a fenestra perilymphaticum on the caudal surface of the labyrinth. Further analysis of other specimens will shed light on this issue and may ultimately argue for yet another independent evolution of a fenestra cochleae.

Discussion

Brain size and behavior

The relationship between endocast volume and body mass has attracted a great deal of attention because it allows a proper assessment of relative brain size (Jerison 1973; Hopson 1977; Hurlburt 1996). Although we can calculate endocranial volume easily and precisely in the digital realm, reliable body masses for extinct animals remain problematic in most cases. This problem is particularly acute in the present case because the specimen from which we have generated a digital endocast (TMP 1989.55.1243) is a disarticulated braincase found in a large bone bed and cannot be reliably associated with any postcranial elements (Currie et al. 2008). Thus, the traditional means of generating a body mass estimate are not available for TMP 1989.55.1243, and we cannot assess relative brain size at the present time. That is, we cannot calculate the encephalization quotient (EQ—Jerison 1973; or its reptilian extension, REQ—Hurlburt 1996), which provides a useful (although somewhat statistically problematic) comparative metric. It may ultimately be possible to establish a rough body mass estimate to accompany a braincase of this size, but until such time, we choose not to introduce poorly constrained estimates into the literature.

Nevertheless, it does not take complicated statistical models to justify the observation that the brain of *P. lakustai* must have been relatively very small in size and basically rudimentary in structure. As noted, the general brain structure was reptilian in that there is no evidence for any of the morphological apomorphies seen in derived theropods and birds. The cerebrum, cerebellum, and optic lobes were not at all expanded. That said, other attributes of *P. lakustai* suggest considerable behavioral sophistication. Its elaborate cranial ornamentations indicate the importance of intraspecific (and maybe also interspecific) behavioral communication and display (Sampson et al. 1997; Sampson 2001). Likewise, the evidence for herding and potentially even migration (Currie 1992) suggests some measure of behavioral complexity. These findings suggest that fairly elaborate behavioral repertoires are consistent with modest to small brain sizes. Moreover, they also put in perspective the relatively expanded brains of such dinosaurs as hadrosaurids, a group whose radiation and behavioral range is often compared with that of ceratopsids. The large brains of many hadrosaurids (Ostrom 1961; Hopson 1979; Evans 2005) may reflect cognitive capabilities lacking in ceratopsids such as *Pachyrhinosaurus*.

The effects of retrodeformation on brain size and form

As discussed, the *P. lakustai* braincase at hand (TMP 1989.55.1243) is somewhat deformed, being laterally compressed. Given that CT scanning digitizes the specimen, the possibility presents itself to address or even correct the fossil's deformation, a process called retrodeformation. A number of different approaches to retrodeformation have been proposed (e.g., Motani 1997; Srivastava and Shah 2006; Angielczyk and Sheets 2007), all of which have their advantages and disadvantages, and all of which require assumptions that we cannot adequately test for TMP 1989.55.1243. Consequently, pending a more formal treatment, we performed a simple transformation in an attempt to improve the cranial endocast to a first approximation. The assumption underlying the transformation is derived from the assessment of this specimen presented by Currie et al. (2008) that the occipital condyle should be about as wide as high. Thus, using Amira's Transform Editor module, we uniformly stretched the transverse dimension until the x dimension of the condyle equaled the y dimension. Thus, we only addressed plastic deformation and ignored brittle deformation (i.e., fractures). Although certainly overly simplistic, this approach to retrodeformation has the advantages of making fewer initial assumptions and being "quick and dirty."

The results returned are generally satisfying (Fig. 5). The braincase itself compares more favorably with less distorted specimens. Moreover, the endocast is credible, agreeing in general proportions with the physical endocasts of *Anchiceratops* (AMNH 5259; Brown 1914) and *Triceratops* (Forster 1996) and our digital endocast of *Centrosaurus*. The retrodeformation has a considerable impact on endocast metrics. The volume of the endocast as preserved (i.e., prior to transformation) is 63.0 cm^3, whereas the retrodeformed volume is 85.4 cm^3, a 35% increase. Even though we have no reliable body mass to accompany these endocranial values, the retrodeformed value would be more appropriate to use in studies of relative brain size. That said, we have chosen to illustrate the nonretrodeformed braincase, endocast, and labyrinth in Figs. 1–4 because these represent the unadulterated documents as preserved in the fossil record.

Sensorineural functional inferences

The dural envelope of *P. lakustai* (as represented by the cranial endocast) was sufficiently "loose-fitting" to the brain, as it was in most non-coelurosaurian dinosaurs, that fine details of brain structure are not recoverable. Nevertheless, based on what is preserved and also the structure of the endosseous labyrinth, some light may be shed on the sensorineural attributes and capabilities of this ceratopsid. Jerison's (1973) "principle of proper mass" (see also Butler and Hodos 2005) affirms that the size of a brain region (e.g., the optic tectum) relates to the biological importance of the function of that region (e.g., processing visual information). Thus, the fact that the olfactory bulbs, cerebrum, optic tecta, and cerebellum are so small (regardless of any retrodeformation) suggests that precise sensory integration and control were not at a premium for *P. lakustai*. Given that some of these regions are enlarged in other dinosaurs (e.g., enlarged olfactory bulbs in sauropods, enlarged cerebrum in hadrosaurids, enlarged optic tecta in many coelurosaurs; Witmer et al. 2008), their small size in *P. lakustai* indicates their lower level of function. One remarkable attribute is the large size of the trigeminal nerve canals. Although some portion of each canal was no doubt occupied by veins, it is reasonable to suggest that the nerves themselves were also enlarged. Nevertheless, it is difficult to know whether tactile sensation (e.g., mechanoreception) was particularly heightened or whether the large size of the nerves simply reflects the enormous head, in that considerable trigeminal innervation would be required for general somatic sensation, as well as for motor supply to the large jaw musculature.

The structure of the inner ear shows that hearing was apparently not a particularly important sensory modality because the cochlea is quite short. The length of the cochlea is directly related to the length of the sensory epithelium (i.e., the basilar membrane; Manley 1990), and cochlear length has long been employed as a measure of hearing capabilities (Wever 1978; Gleich and Manley 2000; Gleich et al. 2004, 2005). The vestibular apparatus, on the other hand, was perhaps a little better developed than in some other neoceratopsians, in that the semicircular canals are somewhat more elongate. This elongation could be linked to a more general sense of balance and equilibrium, but recent research has shown that the semicircular canals have important neural links coordinating eye movements and head turning (Spoor et al. 2007, and references therein; see also Witmer et al. 2008). Thus, *P. lakustai* may have had somewhat better gaze stabilization mechanisms than some other neoceratopsians. We currently have insufficient data to assess whether the slight elongation would be statistically significant, but it is fair to suggest that it would be of doubtful biological significance.

Acknowledgments

We thank Philip Currie for the invitation to work on this project and to contribute to this volume. We thank Eric Snively, Michael Ryan, Philip Currie, and the staff of Canada Diagnostic Centre in Calgary, Alberta, for the CT scanning of the *Pachyrhinosaurus* specimen used here, as well as the *Centrosaurus* specimen mentioned several times. For help with CT scanning of other specimens in the larger study, we thank Heather Rockhold, and O'Bleness Memorial Hospital, Athens, Ohio, as well as Timothy Ryan and Avrami Grader at the Center for Quantitative Imaging, Pennsylvania State University. The manuscript benefited from very useful reviews by Michael Ryan and Hans Larsson. We thank Mark Norell and Carl Mehling (AMNH) for access to their historic collection of dinosaur endocasts, as well as for loan of *Psittacosaurus* and *Protoceratops* specimens. We thank the Royal Tyrrell Museum of Palaeontology for access to the collections and particularly to Darren Tanke for his encyclopedic knowledge of *Pachyrhinosaurus*. The specimen used here represents the first specimen ever jacketed and removed from the field by Hans Larsson, and we thank him for his care. For funding, we acknowledge National Science Foundation grants IBN-0343744 and IOB-0517257 to L.M. Witmer and R.C. Ridgely, the Ohio University College of Osteopathic Medicine, and the Chang Ying-Chien Professorship of Paleontology.

References

Angielczyk, K.D., and Sheets, H.D. 2007. Investigation of simulated tectonic deformation in fossils using geometric morphometrics. Paleobiology, **33:** 125–148.

Brochu, C.A. 2000. A digitally rendered endocast for *Tyrannosaurus rex*. Journal of Vertebrate Paleontology, **20:**1–6.

Brochu, C.A. 2003. Osteology of *Tyrannosaurus rex*: insights from a nearly complete skeleton and high-resolution computed tomographic analysis of the skull. Society of Vertebrate Paleontology Memoir 7, Journal of Vertebrate Paleontology, **22** (Supplement to 2)**:** 1–140.

Brown, B. 1914. *Anchiceratops*, a new genus of horned dinosaurs from the Edmonton Cretaceous of Alberta. With discussion of the origin of the ceratopsian crest and the brain casts of *Anchiceratops* and *Trachodon*. Bulletin of the American Museum of Natural History, **33:** 539–548.

Brown, B., and Schlaikjer, E.M. 1940. The structure and relationships of *Protoceratops*. Annals of the New York Academy of Sciences, **40:** 133–266.

Burckhardt, R. 1892. Das Gehirne von *Triceratops flabellatus* Marsh. Neues Jahrbuch für Mineralogie, Geologie and Paläontologie, **1892:** 71–72.

Butler, A.B., and W. Hodos. 2005. Comparative vertebrate neuroanatomy: evolution and adaptation. 2nd ed. Wiley-Interscience, Hoboken, N.J.

Currie, P.J. 1985. Cranial anatomy of *Stenonychosaurus inequalis* (Saurischia, Theropoda) and its bearing on the origin of birds. Canadian Journal of Earth Sciences, **22:** 1643–1658.

Currie, P.J. 1992. Migrating dinosaurs. *In* The ultimate dinosaur. *Edited by* B. Preiss and R. Silverberg. Bantam Books, New York. pp. 183–195.

Currie, P.J. 1995. New information on the anatomy and relationships of *Dromaeosaurus albertensis* (Dinosauria: Theropoda). Journal of Vertebrate Paleontology, **15:** 576–591.

Currie, P.J., and Zhao, X.-J. 1993. A new troodontid (Dinosauria, Theropoda) braincase from the Dinosaur Park Formation (Campanian) of Alberta. Canadian Journal of Earth Sciences, 30: 2231–2247.

Currie, P.J., Langston, W., Jr., and Tanke, D.H. 2008. A new species of *Pachyrhinosaurus* (Dinosauria, Ceratopsidae) from the Upper Cretaceous of Alberta, Canada. In A new horned dinosaur from an Upper Cretaceous bone bed in Alberta. NRC Research Press, Ottawa, Ontario, Canada. pp. 1–108.

Dodson, P., Forster, C.A., and Sampson, S.D. 2004. Ceratopsia. *In* The Dinosauria. 2nd ed. *Edited by* D.B. Weishampel, P. Dodson, and H. Osmólska. University of California Press, Berkeley, California. pp. 494–513.

Domínguez Alonso, P., Milner, A.C., Ketcham, R.A., Cookson, M.J., and Rowe, T. 2004. The avian nature of the brain and inner ear of *Archaeopteryx*. Nature, **430:** 666–669.

Evans, D.C. 2005. New evidence on brain–endocranial cavity relationships in ornithischian dinosaurs. Acta Palaeontolica Polonica, **50:** 617–622.

Forster, C.A. 1996. New information on the skull of *Triceratops*. Journal of Vertebrate Paleontology, **16:** 246–258.

Franzosa, J.W. 2004. Evolution of the brain in Theropoda (Dinosauria). Ph.D. dissertation, Department of Geological Sciences, University of Texas, Austin, Tex.

Franzosa, J.W., and Rowe, T. 2005. Cranial endocast of the Cretaceous theropod dinosaur *Acrocanthosaurus atokensis*. Journal of Vertebrate Paleontology, **25:** 859–864.

Gilmore, C.W. 1919. A new restoration of *Triceratops*, with notes on the osteology of the genus. Proceedings of the United States National Museum, **55:** 97–112.

Gleich, O., and Manley, G.A. 2000. The hearing organ of birds and Crocodilia. *In* Comparative hearing: birds and reptiles. *Edited by* R.J. Dooling, R.R. Fay, and A.N. Popper. Springer, New York. pp. 70–138.

Gleich, O., Fischer, F.P., Köppl, C., and Manley, G.A. 2004. Hearing organ evolution and specialization: archosaurs. *In* Evolution of the vertebrate auditory system. *Edited by* G.A. Manley, A.N. Popper, and R.R. Fay. Springer, New York. pp. 224–255.

Gleich, O., Dooling, R.J., and Manley, G.A. 2005. Audiogram, body mass, and basilar papilla length: correlations in birds and predictions for extinct archosaurs. Naturwissenschaften, **92:** 595–598.

Hay, O.P. 1909. On the skull and the brain of *Triceratops*, with notes on the brain-cases of *Iguanodon* and *Megalosaurus*. Proceedings of the United States National Museum, **36:** 95–108.

Hopson, J.A. 1977. Relative brain size and behavior in archosaurian reptiles. Annual Review of Ecology and Systematics, **8:** 429–448.

Hopson, J.A. 1979. Paleoneurology. *In* Biology of the Reptilia. Vol. 9. Neurology A. *Edited by* C. Gans. Academic Press, New York. pp. 39–146.

Hurlburt, G.R. 1996. Relative brain size in recent and fossil amniotes: determination and interpretation. Ph.D. dissertation, University of Toronto, Toronto, Ont.

Iwaniuk, A.N., and Nelson, J.E. 2002. Can endocranial volume be used as an estimate of brain size in birds? Canadian Journal of Zoology, **80:** 16–23.

Jerison, H.J. 1973. Evolution of the brain and intelligence. Academic Press, New York.

Kundrát, M. 2007. Avian-like attributes of a virtual brain model of the oviraptorid *Conchoraptor gracilis*. Naturwissenschaften, **94:** 499–504.

Kurochkin, E.N., Saveliev, S.V., Postnov, A.A., Pervushov, E.M., and Popov, E.V. 2006. On the brain of a primitive bird from the Upper Cretaceous of European Russia. Paleontological Journal, **40:** 655–667.

Kurochkin, E.N., Dyke, G.J., Saveliev, S.V., Pervushov, E.M., and Popov, E.V. 2007. A fossil brain from the Cretaceous of European Russia and avian sensory evolution. Biology Letters, **3:** 309–313.

Langston, W., Jr. 1975. The ceratopsian dinosaurs and associated lower vertebrates from the St. Mary River Formation (Maestrichtian) at Scabby Butte, southern Alberta. Canadian Journal of Earth Sciences, **12:** 1576–1608.

Larsson, H.C.E. 2001. Endocranial anatomy of *Carcharodontosaurus saharicus* (Theropoda: Allosauroidea) and its implications for theropod brain evolution. *In* Mesozoic vertebrate life: new research inspired by the paleontology of Philip J. Currie. *Edited by* D.H. Tanke and K. Carpenter. Indiana University Press, Bloomington, Ind. pp. 19–33.

Manley, G.A. 1990. Peripheral hearing mechanisms in reptiles and birds. Springer, New York.
Marsh, O.C. 1896. The dinosaurs of North America. Annual Report of the United States Geological Survey, **16:** 133–244.
Motani, R. 1997. New technique for retrodeforming tectonically deformed fossils, with an example for ichthyosaurian specimens. Lethaia, **30:** 221–228.
Osborn, H.F. 1912. Crania of *Tyrannosaurus* and *Allosaurus*. Memoirs of the American Museum of Natural History, **1:** 1–30.
Osmólska, H. 2004. Evidence on relation of brain to endocranial cavity in oviraptorid dinosaurs. Acta Palaeontologia Polonica, **49:** 321–324.
Ostrom, J.H. 1961. The cranial anatomy of the hadrosaurian dinosaurs of North America. Bulletin of the American Museum of Natural History, **122:** 33–186.
Ridgely, R.C., and Witmer, L.M. 2007. Gross Anatomical Brain Region Approximation (GABRA): a new technique for assessing brain structure in dinosaurs and other fossil archosaurs. Journal of Morphology, **268:** 1124.
Rogers, S.W. 1998. Exploring dinosaur neuropaleobiology: viewpoint computed tomography scanning and analysis of an *Allosaurus fragilis* endocast. Neuron, **21:** 673–679.
Rogers, S.W. 1999. *Allosaurus*, crocodiles, and birds: evolutionary clues from spiral computed tomography of an endocast. Anatomical Record, **257:** 162–173.
Rogers, S.W. 2005. Reconstructing the behaviors of extinct species: an excursion into comparative paleoneurology. American Journal of Medical Genetics, **134A:** 349–356.
Sampson, S.D. 2001. Speculations on the socioecology of ceratopsid dinosaurs (Ornithischia: Neoceratopsia). *In* Mesozoic vertebrate life: new research inspired by the paleontology of Philip J. Currie. *Edited by* D.H. Tanke and K. Carpenter. Indiana University Press, Bloomington, Ind. pp. 263–276.
Sampson, S.D., and Witmer, L.M. 2007. Craniofacial anatomy of *Majungasaurus crenatissimus* (Theropoda: Abelisauridae) from the Late Cretaceous of Madagascar. Society of Vertebrate Paleontology Memoir 8, Journal of Vertebrate Paleontology, **27** (Supplement to 2): 32–102.
Sampson, S.D., Ryan, M.J., and Tanke, D.H. 1997. Craniofacial ontogeny in centrosaurine dinosaurs (Ornithischia: Ceratopsidae): taxonomic and behavioral implications. Biological Journal of the Linnean Society, **121:** 293–337.
Sedlmayr, J.C. 2002. Anatomy, evolution, and functional significance of cephalic vasculature in Archosauria. Ph.D. dissertation, Department of Biological Sciences, Ohio University, Athens, Ohio.

Sereno, P.C., Wilson, J.A., Witmer, L.M., Whitlock, J.A., Maga, A., Ide, O., and Rowe, T. 2007. Structural extremes in a Cretaceous dinosaur [online]. PLoS ONE **2**(11): e1230. doi:10.1371/journal.pone.0001230.

Spoor, F., Garland, T., Jr., Krovitz, G., Ryan, T.M., Silcox, M.T., and Walker, A. 2007. The primate semicircular canal system and locomotion. Proceedings of the National Academy of Sciences, **104:** 10808–10812.

Srivastava, D.C., and Shah, J. 2006. Digital method for strain estimation and retrodeformation of bilaterally symmetric fossils. Geology, **34:** 593–596.

Wever, E.G. 1978. The reptile ear. Its structure and function. Princeton University Press, Princeton, N.J.

Witmer, L.M., and Ridgely, R.C. 2008. The paranasal air sinuses of predatory and armored dinosaurs (Archosauria: Theropoda & Ankylosauria) and their contribution to cephalic architecture. Anatomical Record, 291.

Witmer, L.M., Chatterjee, S., Franzosa, J.W., and Rowe, T. 2003. Neuroanatomy of flying reptiles and implications for flight, posture and behavior. Nature, **425:** 950–953.

Witmer, L.M., Ridgely, R.C., Dufeau, D.L., and Semones, M.C. 2008. Using CT to peer into the past: 3D visualization of the brain and ear regions of birds, crocodiles, and nonavian dinosaurs. *In* Anatomical imaging: towards a new morphology. *Edited by* H. Endo and R. Frey. Springer-Verlag, Tokyo: pp. 67-88.

Zhou, C.-F., Gao, K.-Q., Fox, R.C., and Du, X.-K. 2007. Endocranial morphology of psittacosaurs (Dinosauria: Ceratopsia) based on CT scans of new fossils from the Lower Cretaceous, China. Palaeoworld, **16:** 285–293.